SURFACE IMAGING
for
BIOMEDICAL
APPLICATIONS

T0225435

SURFACE IMAGING
for
BIOMEDICAL
APPLICATIONS

Edited by
Ahmad Fadzil Mohamad Hani
Universiti Teknologi Petronas
Perak, Malaysia

CRC Press
Taylor & Francis Group
Boca Raton London New York

CRC Press is an imprint of the
Taylor & Francis Group, an **informa** business

CRC Press
Taylor & Francis Group
6000 Broken Sound Parkway NW, Suite 300
Boca Raton, FL 33487-2742

First issued in paperback 2017

© 2014 by Taylor & Francis Group, LLC
CRC Press is an imprint of Taylor & Francis Group, an Informa business

No claim to original U.S. Government works

Version Date: 20140114

ISBN 13: 978-1-138-07566-5 (pbk)
ISBN 13: 978-1-4822-1578-6 (hbk)

Library of Congress Cataloging-in-Publication Data

Hani, Ahmad Fadzil Mohamad, author.
  Surface imaging for biomedical applications / Ahmad Fadzil Mohamad Hani.
    p. ; cm.
  Includes bibliographical references and index.
  ISBN 978-1-4822-1578-6 (hardcover : alk. paper)
  I. Title.
  [DNLM: 1. Skin Diseases--diagnosis. 2. Dermoscopy--methods. 3. Image Interpretation, Computer-Assisted--methods. WR 141]

  RL105
  616.5'075--dc23   2013049460

**Visit the Taylor & Francis Web site at
http://www.taylorandfrancis.com**

**and the CRC Press Web site at
http://www.crcpress.com**

The year that I have spent writing and editing this book was made possible with the unwavering support of my research students: Hermawan, Esa, Fitriyah, Nejood, Evan; my colleagues: Dr. Aamir, Dr. Majdi; and my collaborators: Hospital Kuala Lumpur dermatologists: Dr. Azura, Dr. Suraiya, Dr. Felix. Their perseverance and undying search for answers during the course of the research work and clinical studies have led to this piece of work. Thus, I would like to dedicate this work to them for their dedication, perseverance, and patience.

Contents

Preface

As the editor and contributing author, I am motivated and compelled to write and compile this book after several clinical observational studies at Hospital Kuala Lumpur that investigated the use of signal and imaging processing techniques in dermatology for diagnostic and monitoring of skin diseases.

In clinical practice, dermatologists use both visual and tactile inspections to determine types and conditions of skin disorders. These inspection methods are highly subjective and thus require extensive training for use in clinical practice. Since 2004, we have been working with dermatologists in Hospital Kuala Lumpur on various skin disorders in developing objective measurement tools for diagnostic and monitoring purposes. Our measurement tools are developed based on image processing techniques combined with signal and statistical analyses, and are not only objective but also highly accurate.

In this book, we report the development of a psoriasis severity measurement tool called alpha-PASI that performs the Psoriasis Area Severity Index (PASI) gold standard that covers psoriasis lesion erythema, area, thickness, and scaliness. Several signal and image modalities are used. For example, 2D color data is used for determining lesion area, spectrophotometer data for lesion erythema while 3D surface imaging is used to determine thickness and scaliness of lesions.

The physician's global assessment of pigmentary skin disorders such as vitiligo, requires visual inspection by dermatologist, but pigmentation changes due to treatment take three to six months to discern visually. Therapeutic responses of vitiligo treatments are typically very slow and time consuming, and patients respond differently to treatments. Based on skin color model and using advanced image techniques with independent component analysis, the developed VT-Scan enables us to detect minute changes in pigmentation of the skin due to abnormal melanin production, thus reducing the interval between observations to several weeks. With VT-Scan, dermatologists are able to segment vitiligo areas and determine repigmentation areas accurately, allowing the assessment to be conducted within a shorter duration of six weeks compared to the typical three to six months.

The effectiveness of a treatment regime for chronic ulcers can be estimated by measuring changes in the ulcer wound. However, current ulcer management based on visual observation of the ulcer's conditions is not sufficient to determine treatment efficacy. Invasive methods for wound measurements are time consuming and often result in inconsistency of patient care. We have developed a volume ulcer assessment that uses non-invasive 3D imaging techniques to determine ulcer volume objectively. 3D laser scanning (laser triangulation) and optical scanning (structured light) techniques are investigated. Algorithms for solid modelling and volume computation from 3D surface scans are developed.

Most of the diagnosis process carried out by dermatologists for assessment of the acne vulgaris lesion is visual in nature. In grading acne vulgaris, the identification of acne lesion type and the counts for each particular type at various parts of the face, upper back and chest, is a time-consuming and tedious process. As a result, the dermatologists make an estimation of the type and the count of acne lesions and assign the grade based on that approximation; this introduces subjectivity and it also leads to inter- and intra-variability. We have developed an automated system for acne grading that is based on capturing the images of various body parts using the DSLR camera. Contrast enhancement and segmentation are two of the key and essential steps of the automated acne grading system. Contrast enhancement is achieved by artificially generating the high dynamic range images of acne lesions using the local rank transform followed by log-based tone mapping while the acne lesions segmentation method is based on the spectral models of acne lesions.

Lastly, to my fellow biomedical imaging researchers, I hope this book will bring out the passion of research that transcends disciplines and professions.

Ahmad Fadzil Mohamad Hani, FASc, FIEM, PEng, PhD

Centre for Intelligent Signal and Imaging Research (CISIR)
Universiti Teknologi PETRONAS
Seri Iskandar, Perak

Acknowledgments

We would like to thank the Universiti Teknologi PETRONAS staff and ViTrox R&D team for the laboratory and programming support. We acknowledge the various university internal funds and external grants received from the Malaysian Ministry of Science, Technology, and Innovation, and the Ministry of Higher Education for the research work. We would like to thank our collaborators, the dermatologists from the Department of Dermatology, Hospital Kuala Lumpur; many of them played a critical role as principal investigators in the clinical observational studies.

About the Editor

Professor Ahmad Fadzil Mohamad Hani is an expert in the area of image processing and computer vision. He graduated with a BSc (1st Class Honors) in electronic engineering in 1983, and obtained his MSc in telematics in 1984, and his PhD in image processing in 1991 from the University of Essex, UK. He has been actively involved in machine vision and medical imaging research since the early 1990s. His research activities range from fundamental pattern recognition to developing vision applications in the biomedical imaging area such as in retinal vasculature imaging for grading severity of diabetic retinopathy and digital analysis leading to objective assessment for treatment efficacy of ulcer wounds and psoriasis lesions. His current research challenges are developing new analysis techniques for early osteoarthritis and drug addiction using MRI techniques and bio-optics for skin pigmentation analysis. He has authored more than 180 research articles in journals and conference proceedings, been granted several patents, and won several awards for his work. He currently heads the Centre for Intelligent Signal & Imaging Research (CISIR), a university research centre of excellence under the Mission-Oriented Research in Biomedical Engineering at Universiti Teknologi PETRONAS.

Professor Fadzil is a Fellow of the Academy of Sciences, Malaysia and a Fellow of Institution of Engineers, Malaysia. He is a registered professional engineer with the Board of Engineers, Malaysia and a senior member of the Institution of Electrical & Electronic Engineers, Inc. In industry, he sits on the board of directors of ViTrox Corporation Bhd., an R&D and publicly-listed company that develops and manufactures automated vision inspection systems.

1

Skin Surface Roughness Measurement for Assessing Scaliness of Psoriasis Lesions

Ahmad Fadzil Mohamad Hani and Esa Prakasa

CONTENTS

1.1 Introduction

The skin is comprised of several main layers. The epidermis is the outer layer and the dermis at the inner part. The next layer located under the epidermis is the subcutaneous layer. The epidermis is made by the stacking of skin cells: the living cells in the inner layer and the dead cells in the outer layer. In this inner layer the skin cells are produced and matured to replace the dead skin cells. The dead skin cells at the outermost epidermis layer form a tough, flexible, and waterproof mantle on the skin surface. These cells are shed periodically from the skin surface as keratinous scales.

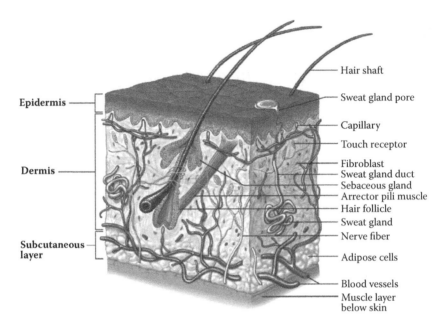

Hair shaft

Sweat gland pore

Capillary

Touch receptor

Fibroblast
Sweat gland duct
Sebaceous gland
Arrector pili muscle
Hair follicle
Sweat gland
Nerve fiber

Adipose cells

Blood vessels
Muscle layer
below skin

Epidermis

Dermis

Subcutaneous
layer

FIGURE 1.1
Human skin is composed of several skin layers [4].

The dermis provides the characteristics of skin such as resistance to tearing and elasticity. This layer contains a thick network of collagen and elastic fibers. Blood vessels (arteries and veins), lymphatics, nerve fibers, connective tissue cells, and immune cells are also compacted in the dermis layer. The deepest layer, the subcutaneous tissue or hypodermis, is not considered a part of human skin. However, this layer provides the protective functions of human skin. The hypodermis is comprised mainly of adipose tissue and some areolar connective tissues. This layer fixes the skin to the body, stores fat reserves, and acts as thermal and mechanical insulation. Excessive amount of adipose tissue can thicken the hypodermis layer. The arrangement of these three main layers—epidermis, dermis, and subcutaneous layer—are depicted in Figure 1.1.

The stratum corneum, which is located in the epidermis layer, is considered to be the outermost layer of the skin structure. It protects the living skin layers from the environment and is constructed of approximately twenty layers of densely packed keratinocytes, the main cell of the epidermis layer, forming 95% of the skin cells [1]. In the inner part, several new keratinocytes are created to replace the older cells in the outer layers. Once created, a keratinocyte will gradually migrate from the inner to the outer layer. With this migration, the old and dead keratinocytes in the outer layer are replaced by the keratinocytes from the inner layers [2]. The keratinocytes transform their shape during the layer migration, from a round shape to a plate shape, in a process known as differentiation. At the end of the skin cell cycle, the dead keratinocytes flake away from skin surface. This final stage is known as the desquamation process. The human body can shed 30,000 to 40,000 dead skin cells from its surface every minute. This shedding equals 4.08 kg of cells annually [3]. The growth process of skin cells, including keratinocytes, occurs through several layers of skin epidermis. The migration process of keratinocytes through the layers within the epidermis is considered the lifetime of a skin cell, which is typically twenty-eight days long.

1.1.1 Psoriasis

Psoriasis is caused by excessive growth of new skin cells. There are several stages of skin cell generation. In the first stage, a new cell is generated from cell division in the lowest layer of the subcutis, that is, the basal layer. In second stage, the cell shape is flattened following cell migration to the outer layers. The cells are now located in the stratum spinosum. This layer contains large cells with spinous profiles. In the third stage, skin cells move to the stratum granulosum above the stratum spinosum. In this stage the cell size becomes smaller compared with its previous size. Lastly, in the fourth stage, the cells perish in the stratum corneum and fall off the skin surface. Figure 1.2 depicts the life stages of a skin cell.

The most common form of psoriasis is psoriasis vulgaris, constituting 80% of all psoriasis cases. In psoriasis vulgaris, scaly papules and plaques are well defined from the surrounding normal skin [8]. For adults, psoriasis is believed to be the most prevalent immune-mediated skin disease, and is

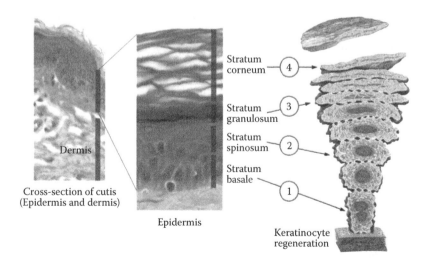

FIGURE 1.2
The stages of skin growing at epidermis layers [5–7].

initiated by an activated cellular immune system. Therefore psoriasis is also considered an organ-specific autoimmune disease [9].

The pathogenesis of psoriasis has not been being completely understood until recently. The immune system and T lymphocytes are considered to be the initiators of psoriasis. In response, the epidermal cell cycle is shortened, resulting in silvery scaly lesions [10]. The psoriasis scales are formed from a number of abnormally stacked cells in the stratum corneum, the outermost surface of the epidermis layer. The granular layer of the epidermis is much reduced by the increase in cell stacking. The clinical features and severities of psoriasis not only vary in time, but also are specific for each individual.

1.1.2 Psoriasis Scaliness

In characterizing psoriasis, four abnormalities are seen. The first abnormality is vascular changes that are shown by dilatation and tortuosity of the papillary blood vessels. At this point there will be elongation and enlargement of the epidermal blood vessels [11]. As a result the skin becomes reddish and this is widely known as psoriasis erythema. The second condition is inflammation, in which polymorphonuclear leukocytes from the dermal vessels move into the epidermis. Here the number of leukocytes of psoriatic lesions significantly increase and many immune-related pathways are activated [11]. In turn, lesions with a large number of activated T helper cells ($CD4^+$ and $CD8^+$) release proinflammatory cytokines. The cytokine is a signaling cell that can activate the body's immune system, resulting in skin inflammation. The third abnormality is hyperproliferation (a rapid rate of cell regeneration) of the keratinocytic layer. This layer protects the body from pathogens, heat, radiation, and water loss. The last abnormality is altered epidermal differentiation. Keratinocytes keep their nuclei in the protected layer (parakeratosis) and lose the granular layer. They grow excessively, with a different appearance compared with normal skin, thus producing skin scales at lesion locations [11]. The epidermal rete is extended and the papillary blood vessels of psoriatic plaques are dilated. The epidermal rete is located at the boundary between the epidermis and dermis layers. The epidermal rete enables the epidermis and dermis to be interlocked into a unified skin layer. These enlarged blood vessels, as shown by psoriatic plaque cross-section in Figure 1.3b, make the affected skin become reddish, which is defined as lesion erythema. Hyperproliferation of keratinocytes is also seen as changes in the stratum corneum. Normally it is thin, but psoriasis causes the layer to become thick and irregular.

Psoriasis lesions typically appear in certain locations, such as the scalp, ears, elbows, umbilicus (belly button), buttocks (gluteal cleft) and genital areas, knees, soles of the feet, fingers, and toes. It is possible for the lesions to grow in moist areas, for example, in the armpits, under the breasts, and the

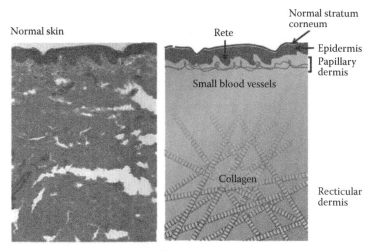

(a) Histology image and corresponding scheme of normal skin

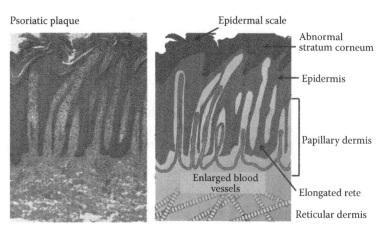

(b) Histology image and corresponding scheme of skin affected by psoriasis

FIGURE 1.3
Skin histology of (a) normal skin and (b) skin affected by psoriasis [9].

groin [12]. Figure 1.4 describes the lesion locations that are commonly found in psoriasis cases.

Psoriasis lesions are typically covered by silvery white scales with varying thicknesses. Plaque psoriasis usually starts in early adulthood and will generally persist for a long time. Plaque psoriasis lesions frequently occur on certain areas, such as the elbows, knees, and scalp [9]. The number of scales may vary among psoriatic patients and also in different areas on the same patient [15]. The lesion is usually clumped as a single skin patch. However, due to improper treatment, the lesion size may be enlarged.

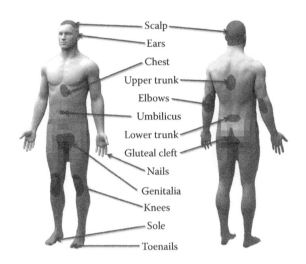

FIGURE 1.4
Common locations of psoriasis lesions growing on human skin. The picture is created by combining graphic materials and description from several references [12–14].

Smaller lesions may merge with neighboring lesions to create a larger lesion.

With more severe psoriasis, the lesion gets thicker with coarse white scales. As indicated in Figure 1.4, psoriasis lesions can affect any body region, particularly the elbows, knees, scalp, palm of the hand, chest, lower back, leg, soles, and nails [16]. Periodic medical treatment for psoriatic patients is important, as the disease cannot be completely cured. Figure 1.5 shows several body regions affected by psoriasis. The images were acquired from data collection sessions in the Dermatology Department, Hospital Kuala Lumpur from 2007 to 2010.

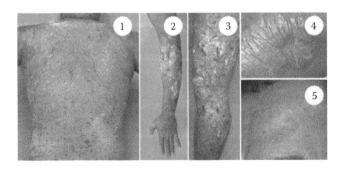

FIGURE 1.5
Psoriasis lesions are found in various body regions: (1) trunk, (2) upper limb, (3) lower limb, (4) elbow, and (5) forehead.

1.2 Assessment of Psoriasis Scaliness

The Psoriasis Area and Severity Index (PASI) scoring method is the gold standard for psoriasis severity assessment [17]. Six clinical psoriasis severity scores—PASI, Body Surface Area (BSA), Psoriasis Global Assessment (PGA), Lattice System Physician's Global Assessment (LS-PGA), Salford Psoriasis Index (SPI), and self-administered PASI (SAPASI)—were systematically reviewed by Bonsard et al. [18]. The PASI scoring is recommended for clinical study. PASI is considered the gold standard because it has been widely applied in clinical studies and is the most validated among the psoriasis scoring methods [18].

The PASI scoring method was introduced by Fredriksson and Pettersson in 1978. The method was proposed to evaluate the clinical efficacy of a new antipsoriatic drug [19]. To determine the total PASI score, four parameters (area [ratio of the lesion area to total body surface area], erythema [color of the lesion inflammation], lesion thickness, and scaliness of the lesion) are required. In the PASI assessment, the human body is divided into four regions: head, trunk, upper limbs and lower limbs. The PASI parameters of the psoriasis lesions are determined for each body region. Dermatologists use their visual and tactile senses to score the PASI parameters. The parameter scores from each region are weighted and totaled to provide a PASI score ranging from 0 to 72. For treatment efficacy, a 75% or greater reduction in the PASI score is considered to be a clinically meaningful improvement [17]. The PASI score is calculated from the following equation:

$$\text{PASI} = 0.1 \times (E_h + T_h + S_h) A_h + 0.2 \times (E_u + T_u + S_u) A_u \\ + 0.3 \times (E_t + T_t + S_t) A_t + 0.4 \times (E_l + T_l + S_l) A_l \tag{1.1}$$

Scores range from 0 to 4 for PASI erythema (E), thickness (T), and scaliness (S), whereas the range for PASI area (A) is from 0 to 6. Four body regions—head, upper limbs, trunk, and lower limbs—are denoted by the subscripts $h, u, t,$ and l, respectively, for each PASI parameter. For PASI scaliness assessments, dermatologists examine several lesions and assign a scaliness score for the examined body region based on the most common score of the lesions. To minimize tediousness in the PASI scaliness assessment, a simplified procedure is performed by dermatologists by selecting a representative lesion for the examined body region. The score obtained from this representative lesion is then considered the scaliness score for the body region.

Although the PASI scoring has been accepted as the gold standard for psoriasis assessment, it is not used in daily practice. PASI scoring is tedious, time-consuming, and subjective. The subjectivity of the scores is influenced by intra- and interrater variation of dermatologists. The reliability and agreement of dermatologist assessments are of concern in many studies [20, 21]. A better assessment method can be decided upon based on the results of agreement analysis.

Therefore an objective and reliable system is required to deal with these problems. An imaging approach is proposed to assess PASI scaliness objectively. To achieve objectivity in the assessment, the PASI scaliness visual descriptors are studied and defined in terms on surface roughness, a measurable feature that can be used to differentiate PASI scaliness scores. Abnormalities of skin can be identified through several skin symptoms, such as itching skin, skin lesions, moles, acne, skin color changes, redness, etc. [22]. Skin lesions can be caused by various diseases. A total of 422 diseases that can cause skin lesions are listed in the RightDiagnosis website [23]. A skin lesion is a superficial growth or patch of the skin that does not appear as normal skin [24]. Skin lesion appearance is specified by visual and tactile descriptions based on primary morphology (size, shape, and thickness), secondary morphology (clustered or distributed), surface texture (roughness), location, and color [25]. Texture determination of normal and abnormal skin is crucial in the field of dermatology measurements, particularly in the evaluation of therapeutic and cosmetic treatments. Skin surface can be characterized by its physical features, such as dryness, smoothness, thickness, and roughness.

Dermatologists conduct a PASI scaliness assessment based on visual descriptors. These descriptors are used as a standard guide for determining PASI scaliness scores. In the proposed approach, surface roughness features are required to represent scaliness severity. The correspondence between surface roughness of the psoriasis lesion and severity needs to be investigated and defined by surface roughness parameters that can be measured from digital surface image data. The current problems of PASI scaliness assessment and surface roughness measurement are as follows:

1. Dermatologist assessment can be subjective due to intra- and inter-rater variability of human assessment. The subjectivity depends on the perception and the clinical experience of the dermatologist.

2. Skin assessment might be performed objectively but the available methods require an invasive treatment on the measured skin. Trained and experienced medical personnel are required to sample and analyze a skin sample using scanning electron microscopy.

3. A noninvasive method can be applied to obtain a high-precision measurement of surface roughness. However, several methods need a skin replica as a representation of the actual skin surface. A skin replica is required because the measuring process uses a precise profilometer (sharp needle or laser beam) to extract the skin surface profile. Although this is a high-precision method, it is not suitable for clinical practice. In addition, skilled personnel are required to build the skin replica and operate the profilometer.

Digital image analysis of skin surface roughness for scaliness classification is a better option to resolve the aforementioned problems. A three-dimensional

(3D) imaging algorithm or clustering algorithm can be applied in the proposed approach. However, there are challenges with this approach as well, including

1. Difficulty in assessing lesion surface roughness on curved skin surfaces. The roughness assessment should be able to measure the vertical deviation of the lesion surface at various points on the lesion and in the various human body regions.
2. To determine vertical deviations at various points on the lesion surface, a reference surface is required to be the zero level of vertical measurement, as commonly applied in industrial applications. However, there are no such surfaces on the skin that can be referred to as the zero level. The reference surface needs to be determined before the vertical deviations can be computed.
3. Due to subjectivity influences on assessment results, a dermatologist's assessment cannot be considered as absolute in evaluating algorithm performance.

Two primary hypotheses are defined as follows:

1. The first hypothesis is that a lesion surface is the superimposed surface between two surfaces—curved and rough surfaces. For surface roughness determination, these surfaces need to be extracted from the lesion surface. Surface roughness of the lesion is calculated by averaging the vertical deviations of the rough surface. Vertical deviation due to the lesion is determined by subtracting the lesion surface from the 3D curved surface at various body regions. The 3D curved surface can be the estimated waviness obtained by fitting a polynomial surface to the lesion surface. Most of the estimated waviness surfaces have the form of curved surfaces.
2. The second hypothesis is that the surface roughness of the lesions can be used to build a clustering algorithm for PASI scaliness scoring. With the availability of large datasets, an unsupervised clustering algorithm can be applied to obtain the boundary levels of lesion surface roughness for PASI scaliness scoring.

Figure 1.6 depicts the cross-sectional view of a skin lesion on normal skin surface. Vertical deviations of the lesion surface can be either positive or negative and the average deviation magnitude is used to represent the surface roughness of the lesion. Here, two objectives are specified:

1. To develop a 3D imaging algorithm for measuring accurately the surface roughness of skin lesions in various body regions.
2. To develop an objective and reliable PASI scaliness scoring system using unsupervised clustering of surface roughness of the skin lesions.

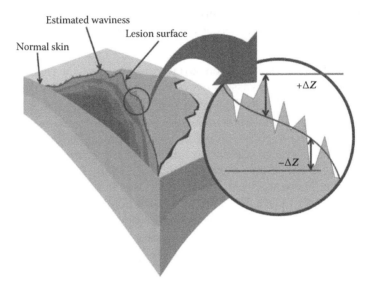

FIGURE 1.6
Cross-sectional view of a skin lesion on normal skin.

1.3 Development of the Surface Roughness Algorithm for Scaliness Assessment

As mentioned earlier, surface roughness is used as a measurable feature to grade psoriasis scaliness. The features are selected based on the surface appearance of the lesion. Rougher surfaces, which are caused by irregular stacks of dead skin cells, are found with more severe psoriasis lesions. Imaging modalities such as 3D optical scanners are now available to perform fast surface scans at high resolution. Figure 1.7 shows the correspondence between the surface roughness of psoriasis lesions and scaliness severity stages. The scores are provided based on the dermatologist's visual and tactile perceptions. In this research, an imaging analysis method is developed to assess scaliness objectively and accurately.

The 3D surface roughness is determined by averaging the vertical deviations of the lesion surface. Because lesions appear on a 3D curved surface (human body), the vertical deviations due to the lesion are determined by subtracting the lesion surface from an estimated waviness surface, as shown in Figure 1.6. A 3D optical scanner with a structured light projection method (the Phase shift Rapid In vivo Measurement of Skin [PRIMOS] portable) is used to acquire lesion surface images. This optical scanner is designed for 3D in vivo measurements of microscopic and macroscopic skin surface structures [26]. The structured light projection method is applied by the PRIMOS camera to obtain a 3D surface image. The method provides a number of advantages, such

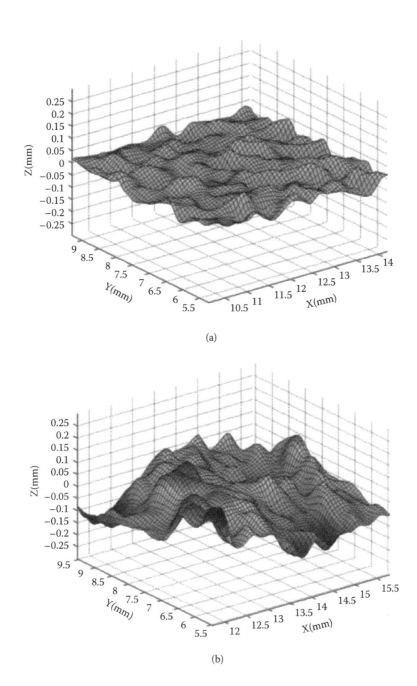

FIGURE 1.7
Psoriasis lesion (a) is scored 1, whereas lesion (b) is scored 4 by a dermatologist based on the assessment of PASI scaliness.

as a standardized capture distance and high-speed scan (<63 ms). High-speed capture is important for skin surface measurements due to the inevitable movements of the subject. PRIMOS has a high-resolution 3D scanner (spatial resolution 0.0062 mm and depth resolution 0.0040 mm), and its image is used as the input to the developed algorithm to determine surface roughness.

A higher-order polynomial surface fit is applied to the rough lesion surface to extract an estimated 3D waviness surface from the rough lesion surface [27]. By subtracting the rough lesion surface from an estimated waviness surface, the vertical deviations of the lesion surface can be exactly determined. The vertical deviations of a lesion's surface are known as the deviation surface. The second- and third-order polynomials are applied, as these polynomial orders are suitable for small surface areas where the vertical undulation of the surface is less.

1.3.1 Polynomial Surface Fitting

Since the lesion surface is not always flat, a polynomial surface fit (second- or third-order polynomials) is required to fit a lesion surface. These polynomial orders are suitable for some small surface areas [28], such as a lesion. In small areas, the vertical undulation of the estimated waviness is less and can be accurately fitted with the second- and third-order polynomials [29]. The general form of polynomials can be written using the following equations [30]:

- The second-order polynomial:

$$z_2(x,y) = (a_1x^2 + a_2x + a_3)\,y^2 + (a_4x^2 + a_5x + a_6)y$$
$$+ (a_7x^2 + a_8x + a_9) \tag{1.2}$$

- The third-order polynomial:

$$z_3(x,y) = (a_1x^3 + a_2x^2 + a_3x + a_4)y^3 + (a_5x^3 + a_6x^2 + a_7x + a_8)y^2$$
$$+ (a_9x^3 + a_{10}x^2 + a_{11}x + a_{12})y$$
$$+ (a_{13}x^3 + a_{14}x^2 + a_{15}x + a_{16}) \tag{1.3}$$

The polynomial coefficients are required for the above second- and third-order polynomials to create a surface. The coefficient of determination (R^2) is used to determine how well the polynomial fits [31]. A good fit can be obtained if R^2 is in the interval [0.9, 1.0]. The best-fitting result of the polynomial order is selected based on the R^2 that is closest to 1. The equation for R^2 is expressed in following equation:

$$R^2 = 1 - \frac{\sum_{i=1}^{M} \sum_{j=1}^{N} (z(x_i, y_j) - w_k(x_i, y_j))^2}{\sum_{i=1}^{M} \sum_{j=1}^{N} (z(x_i, y_j) - \bar{z})^2} \tag{1.4}$$

where $z(x_i,y_j)$ represents the elevation of the lesion surface, $\bar{z}$ represents the elevation average, and $w_k(x_i,y_j)$ is a fitted value at (x_i,y_j) using a kth-order polynomial. The surface roughness is determined by using the average roughness (s_a) equation [32]. In this equation, the surface roughness is calculated by averaging the absolute vertical deviation of all data points. The average roughness, s_a, is defined in equation (1.5). The variable $e(x_i, y_j)$ denotes the vertical deviation of the lesion surface at (x_i, y_j).

$$S_a = \frac{1}{MN} \sum_{i=1}^{M} \sum_{j=1}^{N} |z(x_i y_j) - w_k(x_i, y_j)| = \frac{1}{MN} \sum_{i=1}^{M} \sum_{j=1}^{N} |e(x_i, y_j)| \quad (1.5)$$

1.3.2 Surface Roughness Calculation

The surface roughness algorithm [33] is described in the following steps:

1. Input a 3D lesion surface matrix, $Z_0(x, y)$, with size $M \times N$ (see Figure 1.8a). The coordinates x, y, and its value $z(x, y)$ are then used to calculate some polynomial coefficients. The total data points in the lesion surface is $M \times N$.

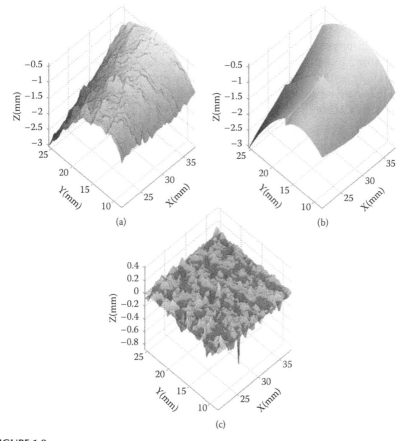

FIGURE 1.8
The 3D surfaces involved in surface roughness determination: (a) lesion surface, (b) estimated waviness, and (c) deviation surface.

2. Divide the lesion surface into 2×2 subdivided surfaces, which produces four subdivided surfaces, namely $D_1, D_2, D_3,$ and D_4. Half division is applied symmetrically to each side of the lesion area, thus the size of each subdivided surface becomes $\frac{M}{2} \times \frac{N}{2} = L$.

3. For each subdivided surface, D_1 to D_4, determine the polynomial coefficients of the selected order from coordinates x, y, and $z(x, y)$ through a matrix inversion. This inversion is performed separately for all subdivided surfaces. Matrix V contains the elements of the polynomial equation, whereas the coefficients of the polynomial equation are stored in matrix A. Matrix Z represents the subdivided lesion surface.

$$VA = ZA = V^{-1}Z \tag{1.6}$$

4. For the first stage, the second-order polynomial is selected to find polynomial coefficient A. Once the coefficients have been determined, the estimated waviness surface can be constructed. The waviness is obtained by applying the calculated polynomial equation with polynomial coefficient A at the evaluated coordinate points (x, y). Equation (1.6) can be written as $V_2 A_2 = Z_{D_1}$ to denote a second-order polynomial fit at subdivided surface D_1.

5. The matrices V_2, A_2, and Z_{D_1} are given in the following forms. Here, variable L refers to the total number of data points at a subdivided surface.

$$V_2 = \begin{bmatrix} x_1^2 y_1^2 & x_1 y_1^2 & y_1^2 & x_1^2 y_1 & x_1 y_1 & y_1 & x_1^2 & x_1 & 1 \\ x_2^2 y_2^2 & x_2 y_2^2 & y_2^2 & x_2^2 y_2 & x_2 y_2 & y_2 & x_2^2 & x_2 & 1 \\ x_3^2 y_3^2 & x_3 y_3^2 & y_3^2 & x_3^2 y_3 & x_3 y_3 & y_3 & x_3^2 & x_3 & 1 \\ x_4^2 y_4^2 & x_4 y_4^2 & y_4^2 & x_4^2 y_4 & x_4 y_4 & y_4 & x_4^2 & x_4 & 1 \\ \cdots & \cdots & \cdots & \cdots & \cdots & \cdots & \cdots & \cdots & \cdots \\ x_L^2 y_L^2 & x_L y_L^2 & y_L^2 & x_L^2 y_L & x_L y_L & y_L & x_L^2 & x_L & 1 \end{bmatrix} \tag{1.7}$$

$$A_2 = [a_1 \quad a_2 \quad a_3 \quad \cdots \quad a_8 \quad a_9]^T \tag{1.8}$$

$$Z_{D_1} = [z_{D_1}(x_1, y_1) \quad z_{D_1}(x_2, y_2) \quad z_{D_1}(x_3, y_3) \cdots z_{D_1}(x_L, y_L)]^T \tag{1.9}$$

6. The matrix of polynomial coefficient A_2 is unknown and inversion $A_2 = V_2^{-1} Z_{D_1}$ is applied to determine its values. The polynomial coefficients of A_2 are determined by applying the equation (1.10).

$$
\begin{bmatrix}
a_1 \\
a_2 \\
a_3 \\
\cdots \\
a_8 \\
a_9
\end{bmatrix}
=
\begin{bmatrix}
x_1^2 y_1^2 & x_1 y_1^2 & y_1^2 & \cdots & x_1^2 & x_1 & 1 \\
x_2^2 y_2^2 & x_2 y_2^2 & y_2^2 & \cdots & x_2^2 & x_2 & 1 \\
x_3^2 y_3^2 & x_3 y_3^2 & y_3^2 & \cdots & x_3^2 & x_3 & 1 \\
x_4^2 y_4^2 & x_4 y_4^2 & y_4^2 & \cdots & x_4^2 & x_4 & 1 \\
\cdots & \cdots & \cdots & \cdots & \cdots & \cdots & \cdots \\
x_L^2 y_L^2 & x_L y_L^2 & y_L^2 & \cdots & x_L^2 & x_L & 1
\end{bmatrix}^{-1}
\times
\begin{bmatrix}
z_{D_1}(x_1, y_1) \\
z_{D_1}(x_2, y_2) \\
z_{D_1}(x_3, y_3) \\
z_{D_1}(x_4, y_4) \\
\cdots \\
z_{D_1}(x_L, y_L)
\end{bmatrix}
$$

(1.10)

7. The waviness surface, $W_2(x, y)$ (Figure 1.8b), is estimated by applying the second-order polynomial surface fitting with coefficient A_2. The estimation is limited for (x, y) the coordinates in D_1 area. An equation for determining $W_2(x, y)$ is shown in the following expressions:

$$W_2 = V_2 A_2 \tag{1.11}$$

$$
\begin{bmatrix}
w_2(x_1, y_1) \\
w_2(x_2, y_2) \\
w_2(x_3, y_3) \\
w_2(x_4, y_4) \\
\cdots \\
w_2(x_L, y_L)
\end{bmatrix}
=
\begin{bmatrix}
x_1^2 y_1^2 & x_1 y_1^2 & y_1^2 & \cdots & x_1^2 & x_1 & 1 \\
x_2^2 y_2^2 & x_2 y_2^2 & y_2^2 & \cdots & x_2^2 & x_2 & 1 \\
x_3^2 y_3^2 & x_3 y_3^2 & y_3^2 & \cdots & x_3^2 & x_3 & 1 \\
x_4^2 y_4^2 & x_4 y_4^2 & y_4^2 & \cdots & x_4^2 & x_4 & 1 \\
\cdots & \cdots & \cdots & \cdots & \cdots & \cdots & \cdots \\
x_L^2 y_L^2 & x_L y_L^2 & y_L^2 & \cdots & x_L^2 & x_L & 1
\end{bmatrix}
\times
\begin{bmatrix}
a_1 \\
a_2 \\
a_3 \\
\cdots \\
a_8 \\
a_9
\end{bmatrix}
$$

(1.12)

8. The deviation surface, $E_2(x, y)$ (Figure 1.8c), is determined by subtracting the estimated waviness, $W_2(x, y)$, from the lesion surface, $Z_{D_1}(x, y)$, as follows:

$$E_2(x, y) = |Z_{D_1}(x, y) - W_2(x, y)| \tag{1.13}$$

9. The coefficient of determination (R^2) is calculated to evaluate the fitting result. Here, $W_2(x, y)$ is accepted if R^2 is within [0.9, 1.0]. The equation for determining R^2 is provided in equation (1.13). The notation

$R_2^2(D_1)$ is used to denote the R^2 of the second-order polynomial surface fitting at the subdivided surface D^1. Equation (1.4) is used to determine this coefficient.

$$R_2^2(D_1) = 1 - \frac{\sum_{i=1}^{\frac{M}{2}} \sum_{j=1}^{\frac{N}{2}} (z_{D_1}(x_i, y_j) - w_2(x_i, y_j))^2}{\sum_{i=1}^{\frac{M}{2}} \sum_{j=1}^{\frac{N}{2}} (z_{D_1}(x_i, y_j) - \overline{z_{D_1}})^2} \tag{1.14}$$

10. Equation (1.15) is used to determine the surface roughness, S_a, at the subdivided surface D_1. The input variables for this equation are the deviation surface $E_2(x, y)$ with matrix size $\frac{M}{2} \times \frac{N}{2}$. The notation $S_{a,2}(D_1)$ is used to represent the surface roughness of the subdivided surface D_1 that is determined by the second-order polynomial surface fitting.

$$S_{a,2}(D_1) = \frac{1}{\left(\frac{M}{2} \times \frac{N}{2}\right)} \sum_{i=1}^{\frac{M}{2}} \sum_{j=1}^{\frac{N}{2}} |z_{D_1}(x_i\, y_j) - w_2(x_i, y_j)| \tag{1.15}$$

11. Repeat steps 4 through 10 to compute the surface roughness values for other subdivided surfaces, D_2 to D_4. These computations are performed separately for each subdivided surface. The overall surface roughness of the lesion surface is obtained by averaging the surface roughness of subdivided surfaces. Here, the surface roughness of a subdivided surface will not be included in final calculation if the R^2 of the polynomial surface fitting is not within an acceptable interval [0.9,1.0]. Thus the overall surface roughness of the lesion surface can be expressed as

$$\overline{S_{a,2}} = \frac{\sum_{i=1}^{N_D} S_{a,2}(D_i)}{N_D}, \quad \text{if} \quad 0.9 \le R_2^2(D_i) \le 1.0, N_D \le 4 \tag{1.16}$$

The overall coefficient of determination R^2 is computed using equation (1.17).

$$\overline{R_2^2} = \frac{\sum_{i=1}^{N_D} R_2^2(D_i)}{N_D}, \quad \text{if} \quad 0.9 \le R_2^2(D_i) \le 1.0, N_D \le 4 \tag{1.17}$$

12. Perform steps 3 to 11, but using the third-order polynomial surface fitting to estimate the surface waviness. For the implementation of a third-order polynomial to the subdivided surface D_1, equation (1.11) is changed to $V_3 A_3 = Z_{D_1}$.

13. The matrix elements of V_3 and A_3 are arranged in the following forms. There is no difference in the input data Z_{D_1} compared with the previous steps. Variable L here represents the total number of data points of the subdivided surface D_1.

$$A_3 = [a_1 \quad a_2 \quad a_3 \quad \cdots \quad a_{15} \quad a_{16}]^T \tag{1.18}$$

14. The inversion of $A_3 = V_3^{-1} Z_{D_1}$ is applied to determine the polynomial coefficient A_3 and its equation is given as follows:

$$
\begin{bmatrix} a_1 \\ a_2 \\ a_3 \\ \cdots \\ a_{15} \\ a_{16} \end{bmatrix} = \begin{bmatrix} x_1^3 y_1^3 & x_1^2 y_1^3 & x_1 y_1^3 & y_1^3 & \cdots & 1 \\ x_2^3 y_2^3 & x_2^2 y_2^3 & x_2 y_2^3 & y_2^3 & \cdots & 1 \\ x_3^3 y_3^3 & x_3^2 y_3^3 & x_3 y_3^3 & y_3^3 & \cdots & 1 \\ x_4^3 y_4^3 & x_4^2 y_4^3 & x_4 y_4^3 & y_4^3 & \cdots & 1 \\ \cdots & \cdots & \cdots & \cdots & \cdots & \cdots \\ x_1^3 y_N^3 & x_N^2 y_N^3 & x_N y_N^3 & y_N^3 & \cdots & 1 \end{bmatrix}^{-1} \times \begin{bmatrix} z_{D_1}(x_1, y_1) \\ z_{D_1}(x_2, y_2) \\ z_{D_1}(x_3, y_3) \\ z_{D_1}(x_4, y_4) \\ \cdots \\ z_{D_1}(x_L, y_L) \end{bmatrix}
$$

$$(1.19)$$

15. The estimated waviness surface $W_3(x, y)$ is obtained by the substitution of coefficient A_3 into equation (1.20).

$$W_3 = V_3 A_3 \tag{1.20}$$

$$
\begin{bmatrix} w_3(x_1, y_1) \\ w_3(x_2, y_2) \\ w_3(x_3, y_3) \\ w_4(x_4, y_4) \\ \cdots \\ w_{16}(x_L, y_L) \end{bmatrix} = \begin{bmatrix} x_1^3 y_1^3 & x_1^2 y_1^3 & x_1 y_1^3 & y_1^3 & \cdots & 1 \\ x_2^3 y_2^3 & x_2^2 y_2^3 & x_2 y_2^3 & y_2^3 & \cdots & 1 \\ x_3^3 y_3^3 & x_3^2 y_3^3 & x_3 y_3^3 & y_3^3 & \cdots & 1 \\ x_4^3 y_4^3 & x_4^2 y_4^3 & x_4 y_4^3 & y_4^3 & \cdots & 1 \\ \cdots & \cdots & \cdots & \cdots & \cdots & \cdots \\ x_1^3 y_N^3 & x_N^2 y_N^3 & x_N y_N^3 & y_N^3 & \cdots & 1 \end{bmatrix} \times \begin{bmatrix} a_1 \\ a_2 \\ a_3 \\ \cdots \\ a_{15} \\ a_{16} \end{bmatrix}
$$

$$(1.21)$$

16. By subtracting the estimated waviness $W_2(x, y)$ from the lesion surface $Z_{D_1}(x, y)$, the deviation surface $E_3(x, y)$ is determined. The equation of this subtraction is $E_3(x, y) = |Z_{D_1}(x, y) - W_3(x, y)|$.

17. The coefficient of the determinant and surface roughness of the measured subdivided surface D_1 are given by following equations:

$$
R_3^2(D_1) = 1 - \frac{\sum_{i=1}^{\frac{M}{2}} \sum_{j=1}^{\frac{N}{2}} (z_{D_1}(x_i, y_j) - w_3(x_i, y_j))^2}{\sum_{i=1}^{\frac{M}{2}} \sum_{j=1}^{\frac{N}{2}} (z_{D_1}(x_i, y_j) - \overline{z_{D_1}})^2} \tag{1.22}
$$

$$
S_{a,3}(D_1) = \frac{1}{\left(\frac{M}{2} \times \frac{N}{2}\right)} \sum_{i=1}^{\frac{M}{2}} \sum_{j=1}^{\frac{N}{2}} |z_{D_1}(x_i, y_j) - w_3(x_i, y_j)| \tag{1.23}
$$

18. Determine $S_{a,3}$ and R_3^2 for the remaining subdivided surfaces, D_2, D_3, and D_4. Hence the overall surface roughness and coefficient of determination are determined by

$$\overline{S_{a,3}} = \frac{\sum_{i=1}^{N_D} S_{a,3}(D_i)}{N_D}, \quad \text{if } 0.9 \leq R_3^2(D_i) \leq 1.0, \, N_D \leq 4 \tag{1.24}$$

$$\overline{R_3^2} = \frac{\sum_{i=1}^{N_D} R_3^2(D_i)}{N_D}, \quad \text{if } 0.9 \leq R_3^2(D_i) \leq 1.0, \, N_D \leq 4 \tag{1.25}$$

19. Coefficients $\overline{R_2^2}$ and $\overline{R_3^2}$ are compared to determine the best order for the polynomial surface fitting. The overall surface roughness $\overline{S_a}$ is obtained from the calculation that gives the $\overline{R^2}$ value closest to 1. For example, if $\overline{R_2^2}$ is found to be closer to 1 than $\overline{R_3^2}$, $\overline{S_{a,2}}$ is considered to be the overall surface roughness. A subtraction between a lesion surface and an estimated waviness is shown in Figure 1.8. This subtraction yields the deviation surface. The average roughness equation is applied to the deviation surface to compute the lesion surface roughness.

1.4 Validation and Implementation of the Scaliness Assessment Based on Surface Roughness

1.4.1 Surface Roughness Measurement: Abrasive Paper

Several abrasive papers with several different roughness grades have been tested to validate the surface roughness measurement on various rough surfaces. There are five grades used for the validation: 16, 24, 60, 80, and 280. Validation on abrasive papers with narrower grade intervals (from 24 to 120) have been presented [34]. The grade value (G) follows the standard of Coated Abrasive Manufacturers Institute (CAMI) [35]. The grade is inversely proportional to the surface roughness. Rougher surfaces have lower grades. The notation $1/G$ is usually used to relate values that are linearly proportional to the surface roughness.

A total of 72 3D surfaces are scanned for each grade. Since there are five grades evaluated in this study, the acquisition gives a total of 280 surfaces. A surface roughness algorithm is applied to the collected 3D surfaces. Table 1.1 lists the surface roughness and its standard deviation for each abrasive grade. The standard deviation represents the measurement precision. As listed in this table, the surface roughness increases proportionally to $1/G$. The average diameter is the average size of the abrading particles embedded in the abrasive paper.

TABLE 1.1

The Surface Roughness of Abrasive Paper

Grade (G)	Average Diameter (mm)	1/G	$\overline{S_a}$ (mm)	σ (mm)	$\overline{S_a} + \sigma$ (mm)	$\overline{S_a} - \sigma$ (mm)
280	0.044	0.004	0.0122	0.0007	0.0115	0.0128
80	0.192	0.013	0.0856	0.0090	0.0766	0.0946
60	0.268	0.017	0.1101	0.0380	0.0722	0.1481
24	0.715	0.042	0.2552	0.0336	0.2216	0.2888
16	1.320	0.063	0.4517	0.0555	0.3962	0.5072

Figure 1.9 shows the 3D surfaces of the abrasive papers. The surface images are ordered from less rough to very rough. Note that impulse noises that appear on the rough surfaces are the result of the reflectivity (specular) of the abrasive particles.

Figure 1.10 shows the plot of surface roughness versus $1/G$ (CAMI grade) for the abrasive papers of Table 1.1. The relation between $1/G$ and S_a is proven by using Pearson's correlation coefficient. The correlation value of S_a and $1/G$ is 0.989, showing a strong relation between them. A positive

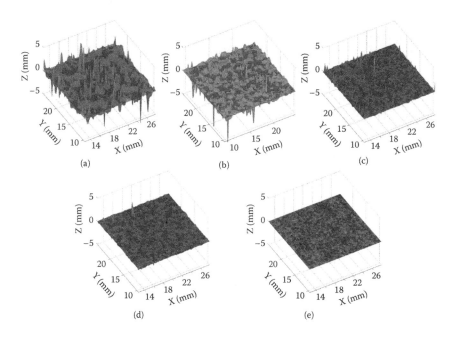

FIGURE 1.9

The 3D surfaces of abrasive papers used in the surface roughness validation. The roughness grades are (a) 16, (b) 24, (c) 60, (d) 80, and (e) 280.

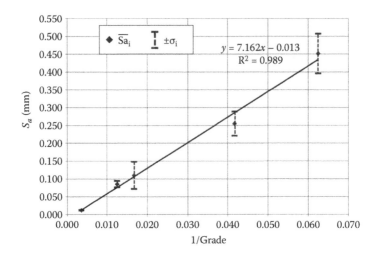

FIGURE 1.10
Plot of surface roughness versus 1/G (CAMI grade) obtained from measurements on abrasive papers.

sign represents a linear relationship, which means that the smoother surface (small S_a value) is represented by a small 1/G. This value proves that the roughness algorithm can be used to determine a roughness grade. Since the surface roughness is an amplitude parameter, its value will relate to a particle size. The rougher surface is composed of particles with a larger diameter and results in a number of higher-amplitude deviations. Therefore 1/G, which is given based on particle size and surface roughness, is linearly correlated.

1.4.2 Surface Roughness on Curve Surface: Mannequin Surface

To model a skin lesion, a medical tape that has regular and uniform texture on its surface is used. It is made from an elastic material and can thus adapt to the curvature of any surface. However, to preserve its texture characteristics, the medical tape has to be pasted onto a smooth surface only.

The average surface roughnesses, $\overline{S_a}$, of thirty-three lesion models are used as the reference. To obtain the average surface roughness reference, these lesion models are pasted on a hard, flat surface. Applying lesion models to a flat surface ensures that only vertical deviations of the lesion model contribute to the roughness calculation. Figure 1.11 shows a lesion model on a flat surface.

The lesion model is scanned three times by applying a series of scans in the scanner mode. The scanning process of thirty-three lesion models gives

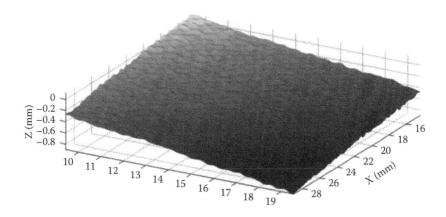

FIGURE 1.11
The 3D surface of a lesion model pasted on a flat surface.

a total of ninety-nine 3D surface images. The surface roughness of a lesion model is represented by the average surface roughness from these three consecutive scans. Additionally, the algorithm is applied separately for each 3D surface to obtain the surface roughness of the lesion models. Since the model preparation is manually performed, it results in a variety of sizes for the lesion model. The smallest and the largest areas are 10.06 mm × 9.93 mm and 14.14 mm × 10.69 mm, respectively. The average surface roughness, $\overline{S}_a$, and its standard deviation are 0.0122 mm ± 0.0011 mm $(\overline{x}_{Ref} \pm \sigma_{Ref})$.

To validate and accuracy of the surface roughness algorithm on the curved surfaces of human skin, similar lesion models are placed on a mannequin used for medical purposes that is made based on the body size of a human adult. A total of 390 lesion models are pasted and distributed onto several locations on the mannequin. The lesion models are placed at the center point of a grid arrangement with size of 40 mm × 30 mm. The center point is selected to provide an accurate image because the highly focused area is located in this point. The region of the lesion models are defined using the four regions (head, upper limbs, trunk, and lower limbs) of PASI scoring. Since each model requires a certain rectangular space, the number of lesions is limited by the available area for each body region. Figure 1.12 shows the lesion model and the mannequin used in the validation study.

The validation process determines the accuracy and precision of a roughness measurement on a 3D curved skin surface. The algorithm is considered valid if the measured surface roughness of the tapes (lesion model) is constant at any location on the skin surface. In this study the surface of the mannequin is used to simulate human skin surfaces. The surface roughness of the lesion models on flat and curved surfaces are then summarized in

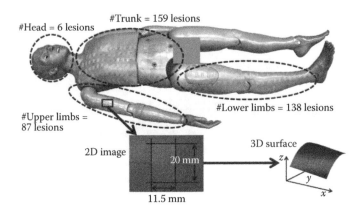

FIGURE 1.12
Validation of surface roughness using lesion models on a life-size mannequin.

histograms, as depicted in Figure 1.13. The average surface roughness, $\overline{S_a}$, and the standard deviation of the lesion models on a curved surfaces were found to be 0.0130 mm ± 0.0013 mm ($\overline{x}_{\text{Curve}} \pm \sigma_{\text{Curve}}$).

1.4.3 Performance Analysis

To determine the accuracy and precision of a surface roughness measurement, the average S_a of the lesion models on a flat surface is used as a reference ($S_{a\text{Ref}}$). The $S_{a\text{Ref}}$ value was found to be 0.0122 mm ± 0.0011 mm ($\overline{x}_{\text{Ref}} \pm \sigma_{\text{Ref}}$). Meanwhile, the S_a of the lesion models on the mannequin surface ($S_{a\text{Curve}}$) was found to be 0.0130 mm ± 0.0013 mm ($\overline{x}_{\text{Curve}} \pm \sigma_{\text{Curve}}$). The following equations are used for the error analysis of the algorithm:

$$\text{Error} = |\overline{x}_{\text{Curve}} - \overline{x}_{\text{Ref}}| \qquad (1.26)$$

$$\text{Precision} = \sigma_{\text{Total}} = \sqrt{\sigma_{\text{Ref}}^2 + \sigma_{\text{Curve}}^2} \qquad (1.27)$$

$$\text{Accuracy} = \left[1 - \frac{|\overline{x}_{\text{Curve}} - \overline{x}_{\text{Ref}}|}{\overline{x}_{\text{Ref}}}\right] \times 100\% \qquad (1.28)$$

From equations (1.26) and (1.27), it is found that the error is |0.0130 − 0.0122|= 0.0008 mm and the precision is $(0.0011^2 + 0.0013^2)^{1/2} = 0.0017$ mm. The algorithm accuracy is computed by applying equation (1.28), which gives $(1 - |0.0130 - 0.0122|/0.0122) \times 100\% = 94.12\%$. In this case, the computed accuracy is the minimum accuracy. It implies that the algorithm accuracy is not less than 94.12%.

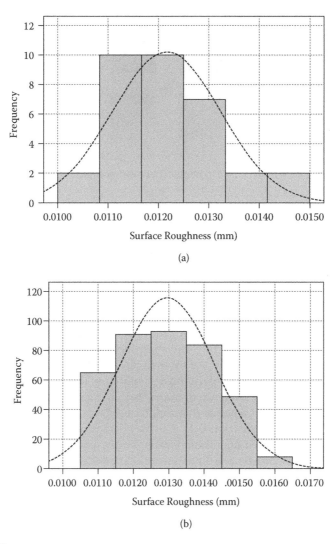

FIGURE 1.13
A histogram of lesion models pasted on (a) flat and (b) curved surfaces.

To satisfy the confidence interval of 99.73%, the observed measurement should be in the range of three standard deviations. By applying this confidence interval, all measurement possibilities can be considered in this analysis. Based on the calculated standard deviation (σ_{Total}), the three standard deviations ($3\sigma_{Total}$) of the algorithm were found to be 0.0051 mm. Thus it can be stated that most of the algorithm measurements (99.73%) will be in the range of ±0.0051 mm. This range value approximates the depth resolution of the PRIMOS camera (0.0040 mm). For vertical deviation determination, the algorithm maintains the precision of the PRIMOS camera. The rotation

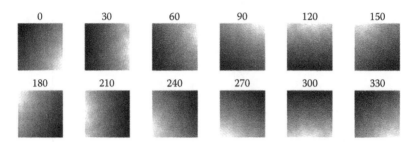

FIGURE 1.14
Rotated surfaces of the lesion model by applying 12 rotation angles.

invariance of the algorithm with the second- and third-order polynomials are tested at the rotated surfaces of the lesion model. A surgical tape on a flat surface with size of more than 40 mm × 30 mm is used as the object. The rotation angles are varied from 0° to 330° with an increment of 30°. Thus twelve rotation angles are applied to evaluate rotation invariance characteristics. Figure 1.14 shows the rotated surfaces of lesion model.

Five measurements are conducted for each angle variation. The S_a of the lesion model with various rotation angles was found to be 0.0128 mm ± 0.0003 mm ($\bar{x} \pm \sigma$). Figure 1.15 shows that the surface roughness values, S_a, for the rotated lesions are within the acceptable accuracy (>95%). The algorithm is therefore considered rotation invariant. This is possible because of the matrix inversion in polynomial coefficients determination does not depend on the element sequence of matrices V and Z. A surface rotation will change the element sequence of these matrices.

To test the rotation invariance of a surface roughness algorithm, a lesion surface obtained in two successive scans is measured. The surface roughnesses of the lesion images are then determined by the same user in a separate

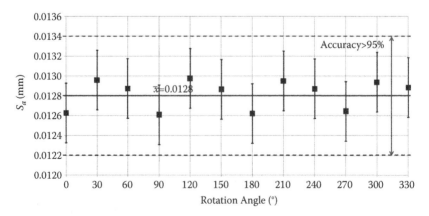

FIGURE 1.15
Surface roughness values of lesion models with variations in rotation angle.

calculation. Here, 465 samples of the lesions (930 images) are used. The lesion surface between the first and second scan is made slightly different.

To prove that the algorithm is invariant to the rotation of the measured surface, the repeatability of the successive scans with rotation angles is evaluated. The definition of repeatability is a closeness between independent results that are obtained through the same method on an identical object, same operator, same conditions, and performed in a short time interval [36]. The absolute differences between the two measurements are used to determine the system repeatability. The repeatability itself can be accepted if 95% of the absolute measurement differences are less than two standard deviations of the measurement difference ($2\sigma_{Diff}$), as suggested by Bland and Altman [37].

From successive measurements, the standard deviation of measurement differences σ_{Diff} was found to be 0.011 mm. Since 95.27% of the measurement differences (443 of 465 lesions) are within the acceptable repeatability ($2\sigma_{Diff} = 0.021$ mm), the system repeatability of the surface roughness algorithm can be accepted. Further analysis showed that the measurement differences are due to manual segmentation of the lesion region of interest (ROI), resulting in differences in the ROI area sampled between assessments. This problem occurs in a segmentation process, particularly in the case of large lesions.

1.4.4 Classification of Lesion Roughness for Scaliness Assessment

Data classification is differentiated into two approaches: supervised and unsupervised. As mentioned by Jain et al. [38], supervised classification is referred to as discriminant analysis, whereas clustering is considered to be an unsupervised classification. For the supervised classification, the training data are labelled and used to define the cluster descriptions. These descriptions are then used to classify the inputted data. In contrast, the unsupervised classification uses unlabelled data for the training. The data are clustered based on their similarity criteria. The process is performed iteratively until the centroids are maximally separated [38]. In data management, clustering algorithms are required due to the increasing volume and variety of data. Clustering algorithms enable users to understand, process, and summarize large amounts of data [39]. By observing the clustered data, analysis and decision making caFuzzy c-means (FCM) clustering is a type of unsupervised clustering algorithm that has been widely applied. Many research works have applied FCM in segmentation applications that relate with skin features. Leung et al. [40] applied FCM clustering to the lip from face skin. Color and shape information are used to classify lip from non-lip regions. Chahir and Elmoataz [41] applied FCM clustering to differentiate and segment skin regions. Color information is selected as the classification feature. The algorithm is useful for data mining purposes. Zhou et al. [42] proposed an improved FCM, anisotropic mean shift based fuzzy c-means (AMSFCM). It is used to find the lesion boundary of malignant melanoma in dermoscopy images. Comparing the AMSFCM with the FCM algorithm, the sensitivity increases from 0.739 (FCM) to 0.776 (AMSFCM). However, the specificity values (0.99) are found to be same for both algorithms. Cucchiara and Grana [43] implemented an FCM clustering algorithm that is applied sequentially. Initially the FCM clustering is used to segment the lesion from the normal skin. In the next stage, the same clustering method is applied to the lesion region. From this subsequent clustering, the internal structure of the lesion can be categorized into several types. In this chapter, FCM clustering algorithms are applied to improve scoring performance. Soft decisions arising from FCM are similar to the fuzziness of nature.

FCM clustering was initially proposed by Bezdek et al. in 1984 [44]. The algorithm has been developed based on a fuzzy set proposed by Zadeh [45]. Bezdek et al. introduced this algorithm in order to improve the k-means algorithm. FCM minimizes the limitations of the hard clustering algorithm, e.g. k-means clustering, by associating each data point to all existing clusters with a membership degree [46]. The membership degree is represented by a membership matrix, $U = [u_{ik}]$. The membership degree of a data point in a particular cluster can be considered the probability of this data point belonging to the cluster. Subscripts i and k are used to index the data point and the cluster, respectively. The membership degrees can be any real values, but it must be between 0 and 1. A data point is associated to each cluster and its belonging is determined by the membership degrees of the clusters. The accumulation of these degrees is always equal to 1 [47].

Membership degrees of the training dataset are obtained from the iteration of FCM clustering. Here, the size of the training dataset given to the FCM algorithm is 1351 3D lesion surfaces. More than 90% of the training dataset is acquired from a single assessment of the psoriasis lesions. Another dataset is obtained from double assessments, but the data cannot be included because it is used as the testing dataset. The degrees are clustered and scattered according to the four clusters of scaliness scores, as depicted in Figure 1.16.

The Gaussian fitting function is used to obtain the membership degrees or the probabilities of surface roughness at any input value. The general form of the Gaussian equation for representing the membership function of scaliness score is given by equation (1.29). The coefficients of scaliness score functions are listed in Table 1.2. The membership degrees of clustered data have been fitted to the Gaussian function with $R^2 \approx 1$. The membership degree of a certain surface roughness, S_a, is calculated by applying all membership functions, $P_n(S_a)$, from $n = 1$ to $n = 4$. The subscript n on variable $P_n(S_a)$ represents the group of scaliness score. Equation (1.29) is constructed from the summation of two Gaussian fitting functions, which have different means and standard deviations. For the Gaussian function, coefficient a_1 is the maximum height of the distribution, b_1 is the mean of the Gaussian distribution, and c_1 is the standard deviation.

$$P_n(S_a) = a_1 \exp\left(-\left(\frac{S_a - b_1}{c_1}\right)^2\right) + a_2 \exp\left(-\left(\frac{S_a - b_2}{c_2}\right)^2\right) \tag{1.29}$$

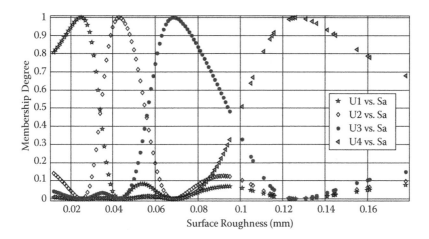

FIGURE 1.16
Membership degrees scattering of a clustered dataset.

The membership functions of PASI scaliness scores are plotted in Figure 1.17. The PASI scaliness score is determined by applying the rule of scaliness score as written in equation (1.30). A surface roughness value is classified into a particular score if the value gives the highest membership degree (probability) among the scaliness scores.

$$S(S_a) = \begin{cases} 1, & max(P_1(S_a), P_2(S_a), P_3(S_a), P_4(S_a)) = P_1(S_a) \\ 2, & max(P_1(S_a), P_2(S_a), P_3(S_a), P_4(S_a)) = P_2(S_a) \\ 3, & max(P_1(S_a), P_2(S_a), P_3(S_a), P_4(S_a)) = P_3(S_a) \\ 4, & max(P_1(S_a), P_2(S_a), P_3(S_a), P_4(S_a)) = P_4(S_a) \end{cases} \tag{1.30}$$

Some examples of scaliness score calculations are listed in Table 1.3. Sample 1, a lesion with a surface roughness of 0.039 mm would be classified with a PASI scaliness score of 2. Membership degrees of the sample for score clusters 1 through 4 are 0.100, 0.913, 0.005, and 0.000, respectively. The highest

TABLE 1.2

Coefficients of Gaussian Functions for the Roughness Classification of PASI Scaliness Scores

Scaliness score	a_1	b_1	c_1	a_2	b_2	c_2	R^2
Score 1 (S_1)	0.609	0.029	0.007	0.888	0.018	0.011	0.994
Score 2 (S_2)	0.514	0.038	0.006	0.858	0.048	0.010	0.992
Score 3 (S_3)	0.635	0.064	0.010	0.784	0.081	0.018	0.995
Score 4 (S_4)	0.708	0.118	0.023	0.792	0.160	0.039	0.997

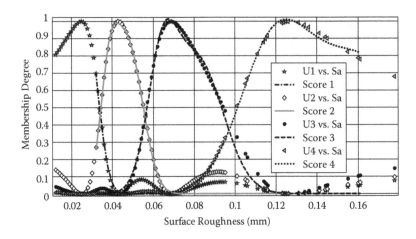

FIGURE 1.17
Membership functions of PASI scaliness scores.

degree is obtained at the second cluster, $P_2(\overline{S}_a) = 0.913$, therefore the sample can be classified as score 2.

The FCM algorithm is evaluated by applying a test dataset. The dataset is obtained from two successive scans of lesion surfaces. From 324 data points tested, a total of 295 (91.0 %) data points were correctly classified in the double assessment sessions. Figure 1.18 describes the complete scaliness scores that have been classified by applying the FCM algorithm. Misclassified data points were found for twenty-nine (9.0 %) data points. These occur at locations near the boundary levels of the score groups. Here the boundary levels of the FCM algorithm would not be considered in the classification process. These boundary levels are determined in order to find the transition boundary from a certain score to another score group.

A boundary level is defined as the intersection point of two overlapping membership functions. Figure 1.19 shows an example of a boundary level between score 1 and score 2. The boundary level is depicted as a gray circle.

TABLE 1.3

Calculations of the PASI Scaliness Scores by Applying the FCM Algorithm

Cluster	$\overline{S}_a$ (mm)	$P_1(\overline{S}_a)$	$P_2(\overline{S}_a)$	$P_3(\overline{S}_a)$	$P_4(\overline{S}_a)$	P_{max}	Score
1	0.039	0.100	0.913	0.005	0.000	0.913	2
2	0.017	0.908	0.000	0.000	0.000	0.908	1
3	0.063	0.000	0.085	0.911	0.004	0.911	3
4	0.112	0.000	0.000	0.044	0.837	0.837	4

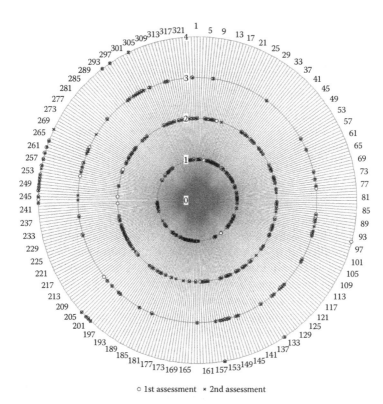

○ 1st assessment × 2nd assessment

FIGURE 1.18
Clustering results of the FCM algorithm on the test dataset.

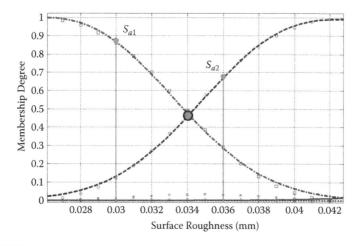

FIGURE 1.19
A boundary level (graf circle) splits membership functions of score 1 (dot-dashed line) and score 2 (dashed line).

TABLE 1.4

Average of Euclidean Distances of Surface Roughness and Boundary Levels of the FCM Classification Algorithm

Misclassification Type	$\overline{D_{B_1}}$ (mm)	$\overline{D_{B_2}}$ (mm)	$\overline{D_{B_3}}$ (mm)	N
Type 1, score 1 and 2	0.003	0.030	0.088	17
Type 2, score 2 and 3	0.033	0.004	0.056	8
Type 3, score 3 and 4	0.091	0.060	0.004	4

At this point both overlapping membership functions (score 1 and score 2) give the same membership degree for a certain surface roughness. This membership degree relation can be expressed by P_1 (0.034) = P_2 (0.034) = 0.4564. A misclassification case appears when the measured surface roughness values are located in the left and the right sides of the boundary level. The first assessment, S_{a1}, is classified as score 1 because the membership function of score 1 gives the largest membership degree at P_1 (0.030) = 0.8684. In the second assessment, the highest membership degree is P_2 (0.036) = 0.6781. Thus the measured surface roughness, S_{a2}, is considered as score 2. Finally, all boundary levels of the FCM algorithm are obtained from the intersection point of the membership functions. The boundary levels of misclassification types 1, 2, and 3 are 0.034 mm, 0.056 mm, and 0.097 mm, respectively. In the FCM algorithm, the boundary levels depend on the curve shape of the membership functions and are not exactly located at the midpoint of two cluster centroids.

The average Euclidean distance is computed to find the closeness of the measured surface roughness with the boundary levels. Table 1.4 shows the closeness of the measured data to the boundary levels of the FCM algorithm. The data are summarized from misclassified data points. Misclassification type 1 exists when the measurement results are located near the first boundary level, B_1. The average distance $\overline{D_{B_1}}$ (0.003 mm) is small and not more than 10% of D_{B_2} (0.030 mm) and D_{B_3} (0.088 mm). Misclassification types 2 and 3 are also indicated by the closeness of the measured surface roughness to the boundary levels. A small value of D_{B_2} (0.004 mm) shows the closeness of the measured surface roughness to the boundary level of misclassification type 2. The occurrence of misclassification type 3 is explained by a small distance value of D_{B_3} (0.004 mm).

These small distances do not occur for the cases with correct classifications. The measured surface roughnesses are located far from the boundary levels. Therefore misclassification cases could be avoided in these measurements. Here, the averages of the Euclidean distances are also found to be greater than $3\sigma_{Total}$ (0.005 mm). The distances for all score groups are summarized in Table 1.5.

Two methods are recommended to minimize misclassification cases. The first method is performed by acquiring more data points if the surface

TABLE 1.5

Average Euclidean Distances of Surface Roughness and Boundary Levels Determined from Correct Classification Cases of the FCM Algorithm

Score Group	$\overline{D_{B_1}}$ (mm)	$\overline{D_{B_2}}$ (mm)	$\overline{D_{B_3}}$ (mm)	N
Score 1	0.012	0.043	0.101	104
Score 2	0.014	0.018	0.076	114
Score 3	0.053	0.022	0.036	55
Score 4	0.128	0.097	0.039	22

roughness lies close to the boundary levels. An average value of this multiple measurement is then applied to score the surface roughness. Table 1.6 summarizes the surface roughness ranges that need to be considered for performing multiple measurements.

In the second method, the score group is determined by comparing the membership degrees of the decided score groups. As exemplified in Figure 1.19, the final score cannot be determined because there are two different scores from two measurement sessions. The final score can be obtained by comparing the maximum membership degrees of the first (P_1 (0.030) = 0.8684) and the second (P_2(0.036) = 0.6781) assessments. From the comparison, it is known that 0.8684 > 0.6781, therefore the final score for this data point is score 1. A mathematical expression is formulated to determine the final score $S(S_a)$ from n-times measurements. Let $m_i(S_i, P_{max,i})$ represents the ith measurement, which has a final score S_i at a maximum membership degree $P_{max,i}$. Then, the equation can be given by

$$S(S_a) = S_k, \ \max(P_{max,1}, P_{max,2}, \ldots, P_{max,k}, \ldots, P_{max,n}) = P_{max,k}(S_a). \quad (1.31)$$

1.4.5 Agreement Analysis of the Scaliness Assessment

The interrater variation of two independent observers can be evaluated by using κ coefficient analysis. The κ coefficient was proposed by Cohen in 1960

TABLE 1.6

The Surface Roughness Ranges Near the Boundary Levels of the FCM Classification Algorithm

Misclassification Type	Surface Roughness Interval (mm), $S_a - 3\sigma_{Total} \leq B_i \leq S_a + 3\sigma_{Total}$
Type 1, score 1 and 2	$0.029 \leq S_a \leq 0.084$
Type 2, score 2 and 3	$0.051 \leq S_a \leq 0.106$
Type 3, score 3 and 4	$0.092 \leq S_a \leq 0.147$

and has been widely used to measure agreement among clinicians on the scores of a medical assessment. The maximum value ($\kappa = 1.0$) represents that the observers agree on all of examination samples. Landis [48] interprets the agreement coefficient into various grades. For example, the κ intervals 0.61 to 0.80 and 0.81 to 0.99 are categorized as substantial agreement and almost perfect agreement, respectively.

To evaluate agreement between two dermatologists, the κ coefficient value of the PASI scaliness scores of 1283 lesions are determined. The κ coefficient value between dermatologist 1 and dermatologist 2 is 0.55 and is categorized as moderate agreement. The κ coefficient of 0.55 indicates that the dermatologists have achieved 55% agreement of the total assessment and 45% agreement would be expected by chance. Perfect agreement is achieved if the κ coefficient is greater than 0.80. As a result, the dermatologists' scores are not considered as reliable in evaluating the algorithm performance since the κ coefficient among the dermatologists is only 0.55.

To evaluate the performance of the PASI scaliness algorithm, several lesion samples are imaged in two successive scans. The PASI scaliness scores of the lesion images are then analyzed by the same user in separate calculations. The number of samples is 324 lesions (648 images). The number of tested lesions is less than the number of lesions scored by the dermatologists because not all the lesions were scanned twice. The κ coefficients between the first assessment and the second assessment that are obtained from the FCM algorithm are evaluated. Tables 1.7 and 1.8 show the agreement summary of the dermatologists and FCM clustering, respectively. Table 1.9 compares the κ coefficients obtained from the dermatologist assessment and FCM clustering. Perfect agreement between the first and second assessments for the FCM algorithm is achieved since the κ coefficient (0.8708) is found to be greater than 0.81. FCM clustering, as a soft clustering, can better solve the subjectivity problems of dermatologist assessments.

TABLE 1.7

Summary of Agreement of Scoring by Dermatologists

First Assessment vs. Second Assessment		Scores of Dermatologist 2 Assessment				Total
		1	2	3	4	
Scores of dermatologist 1 assessment	1	634	200	4	0	838
	2	37	204	39	0	280
	3	4	24	75	5	108
	4	1	0	16	40	57
Total		676	428	134	45	1,283

TABLE 1.8

Summary of Agreement of Scoring by Applying the FCM Algorithm

First Assessment vs. Second Assessment		Scores of Second Assessment				Total
		1	2	3	4	
Scores of First Assessment	1	104	12	0	0	116
	2	5	108	4	0	117
	3	0	6	59	1	66
	4	0	0	1	24	25
Total		109	126	64	25	295

TABLE 1.9

κ Coefficients of Dermatologist Assessment and FCM Clustering

Assessment Method	κ Coefficient	N
Dermatologist	0.5500	1283
FCM	0.8708	324

1.5 Conclusion

This chapter addresses the problem of assessing lesion scaliness, particularly for psoriasis lesions. Visual inspections using scaliness assessments are neither accurate (objective) nor consistent (reliable). The problem was formulated into a surface imaging algorithm where lesion scaliness is regarded as surface roughness of the lesion.

This chapter details the development of a surface roughness algorithm for assessing the scaliness of psoriasis lesions. A 3D optical scanner is used to acquire lesion surface images. The estimated 3D waviness surface is extracted from the rough lesion surface by applying a high-order polynomial surface fitting. The rough lesion surface is subtracted from the estimated waviness to obtain the vertical deviations of the lesion surface. Surface roughness is then calculated by averaging the absolute values of the vertical deviations.

The algorithm has been validated on both a standardized rough surface and curved surfaces. This validation was conducted in order to determine the accuracy and precision of the developed algorithm. A set of abrasive papers with different roughness grades were evaluated to perform algorithm validation on the standardized rough surface. From the validation it was found that the measured surface roughness of the developed algorithm is linearly related to the standardized roughness grade. Medical tape is used to model skin lesions since its elasticity can adapt to the surface curvature. The lesion models are placed on the curved surfaces of a life-size medical mannequin. The validation process calculates the accuracy and precision of the roughness measurement on 3D curved skin surfaces. The algorithm was considered valid if the surface roughness of the lesion model was constant at any location. From the analysis, it was found that the algorithm accuracy and precision are high. The surface roughness algorithm was also tested at varying rotation angles. The results prove the rotation invariance characteristic of the surface roughness algorithm. The rotation invariance of the surface roughness algorithm was also tested on psoriasis lesion surfaces obtained from two successive scans. These successive scans are slightly different due to rotation and translation. From the measured surface roughness, it was found that the system repeatability is acceptable. From these findings, it can be shown that the developed algorithm has been validated at a significant level.

An unsupervised clustering method, FCM clustering, was applied to score psoriasis scaliness. There are four clusters representing four sets of PASI scaliness scores. The FCM clustering iteration provides the membership degrees of each data point. The Gaussian fitting function was fitted to the clustered dataset in order to obtain the membership degrees of the dataset. Meanwhile, the PASI scaliness score was determined by comparing the membership degrees of input data to the existing clusters. The input surface roughness is classified into a particular score cluster if it has the highest membership degree among the clusters of scaliness scores. κ coefficient analysis was used to evaluate the interrater variation of two independent observers. The κ coefficient was found to be less than 0.81. Therefore the dermatologist assessment is not considered reliable for evaluating algorithm performance. The agreement analysis of the PASI scaliness algorithm was performed by comparing two assessments of successive scans. The κ coefficient between the double assessments of the FCM clustering algorithms was found to be greater than 0.81 and thus is considered as a perfect agreement. These remarkable results provide a high level of confidence for implementing the developed system for objective clinical assessment.

Appendix: MATLAB code

```
% Function hdrload() is created by Jeff Daniels (Dec 20, 2002). The file
% is obtained from http://www.mathworks.com/matlabcentral/fileexchange/2973
% -mhdrload
-m
% ====================================================================
% ========== OPEN A 3D SURFACE FILE ==================================
% ====================================================================
close all; clear all; clc;
```

```
% Type the folder location in the following line
% folderToBeOpened = 'K:\3D mannequins - with lesions\male\MALE.3TRUNKok\60\';

cd(folderToBeOpened);
[txtFileNm, folderToBeOpened] = uigetfile('*.txt','Select the M-file');
% Convert the file extention from "txt" to "jpg". The jpg file will be
% opened to segment the lesion area
N = length(txtFileNm);
imgFileNm = txtFileNm;
imgFileNm(N-2:N)='jpg';
% ==================================================================
% ==================================================================


% =============================================
% ========== READ 2D IMAGE AND SEGMENT THE LESION AREA ========
% =============================================
I2d = imread([folderToBeOpened,imgFileNm]);
[m2D,n2D] = size(I2d);
n2D = n2D/3;
% Rotate the 2D image coordinate. The objective is to make it
% corresponds with the 3D surface coordinate
for iii=1:m2D
forjjj=1:n2D
id = m2D - iii + 1;
I2dT(id,jjj,:) = I2d(iii,jjj,:);
end;
end;
I2d = I2dT;
clear I2dT;
figNameStr = ['File Name ',folderToBeOpened,imgFileNm];
figure('Name',figNameStr);

% Show the 2D image to enable user performs manual segementation
imshow(I2d);
[BW,x,y] = roipoly;
close(figNameStr);

% Set the boundary points (top-left and bottom-right) of
% the measured lesion surface
minX = round(min(min(x))); maxX = round(max(max(x)));
minY = round(min(min(y))); maxY = round(max(max(y)));
% ========================================================================
% ========================================================================


% ========================================================================
% ========== READ TXT FILE AND CONVERT IT INTO A SURFACE MATRIX ==========
% ========================================================================
[hdr, A] = hdrload([folderToBeOpened,txtFileNm]);
initX = A(1,1);
newX = 1e10;
N = 1;
whileinitX ~= newX
    N = N + 1;
newX = A(N,1);
end;
N = N - 1;
lengthA = length(A);
M = lengthA/N;
```

```
% Create a grid of X coordinate system
xArr = A(1:N,1);
for i=1:M
I.X(i,:) = xArr;
end;
% Create a grid of Y coordinate system
for i=1:M
idx = 1 + (i-1)*N;
yArr(i) = A(idx,2);
end;
for j=1:N
I.Y(:,j) = yArr;
end;
% Create a grid of Z coordinate system
idxA = 0;
for i=1:M
for j=1:N
idxA = idxA + 1;
I.Z(i,j) = A(idxA,3);
end;
end
% ========================================================================
% ========================================================================


% ========================================================================
% ========== ROI EXTRACTION FROM THE 3D SURFACE ==========================
% ========================================================================
k = 1; % k = 1 is used for the highest image resolution [480 x 640 pixels]
minXSc = round(k*minX);
maxXSc = round(k*maxX);

minYSc = round(k*minY);
maxYSc = round(k*maxY);

Xcr=I.X(minYSc:maxYSc, minXSc:maxXSc);
Ycr=I.Y(minYSc:maxYSc, minXSc:maxXSc);
Zcr=I.Z(minYSc:maxYSc, minXSc:maxXSc);

I2.X = Xcr;
I2.Y = Ycr;
I2.Z = Zcr;
% ========================================================================
% ========================================================================


% ========================================================================
% ========== DIVIDE THE SAMPLE INTO 4 SUB-DIVIDED SURFACES ===============
% ========================================================================
noDiv = 2;
[M,N] = size(I2.Z);

mD = floor(M/noDiv);
nD = floor(N/noDiv);

for i=1:(noDiv)
cI(i) = 1 + (i-1)*mD;
cJ(i) = 1 + (i-1)*nD;
end;
```

```
idxSd = 0;
for i=1:(noDiv)
for j=1:(noDiv)
idxSd = idxSd + 1;
        I2sDiv(idxSd).X = I2.X(cI(i):(cI(i)+mD-1),cJ(j):(cJ(j)+nD-1));
        I2sDiv(idxSd).Y = I2.Y(cI(i):(cI(i)+mD-1),cJ(j):(cJ(j)+nD-1));
        I2sDiv(idxSd).Z = I2.Z(cI(i):(cI(i)+mD-1),cJ(j):(cJ(j)+nD-1));
end;
end;
% =========================================================================
% =========================================================================


% =========================================================================
% ========= APPLY 2nd ORDER SURFACE FITTING TO EXTRACT ROUGHNESS =========
% =========================================================================
noOfAcceptedFitting = 0;
RaAcc = 0;
RsqAcc = 0;
foriSd=1:(noDiv^2)
    [M,N] = size(I2sDiv(iSd).Z);

    % Copy the data of subdivided surfaces into X, Y, and Z matrices
    X = I2sDiv(iSd).X(1:M,1:N);
    Y = I2sDiv(iSd).Y(1:M,1:N);
    Z = I2sDiv(iSd).Z(1:M,1:N);

    % Create matrices required for determining matrix coefficients
rowIdx = 0;
for i=1:M %noDivSp
for j=1:N %noDivSp
rowIdx = rowIdx + 1;
            Xv = X(i,j);
Yv = Y(i,j);
Zv = Z(i,j);

nYv = 2;
V(rowIdx,1) = (Xv^2)*(Yv^nYv);
V(rowIdx,2) = (Xv^1)*(Yv^nYv);
V(rowIdx,3) = (Xv^0)*(Yv^nYv);

nYv = 1;
V(rowIdx,4) = (Xv^2)*(Yv^nYv);
V(rowIdx,5) = (Xv^1)*(Yv^nYv);
V(rowIdx,6) = (Xv^0)*(Yv^nYv);

nYv = 0;
V(rowIdx,7) = (Xv^2)*(Yv^nYv);
V(rowIdx,8) = (Xv^1)*(Yv^nYv);
V(rowIdx,9) = (Xv^0)*(Yv^nYv);

Q(rowIdx,1) = Zv;
end;
end;
    % Find matrix coefficients by applying matrix inversion
VtV= V'*V;
cf = pinv(VtV)*V'*Q;

    % Create estimated waviness based on calculated matrix coefficients
```

```
Zcc = 0;
nYv = 2;
Zcc = Zcc + cf(1)*(X.^2).*(Y.^nYv);
Zcc = Zcc + cf(2)*(X.^1).*(Y.^nYv);
Zcc = Zcc + cf(3)*(X.^0).*(Y.^nYv);


nYv = 1;
Zcc = Zcc + cf(4)*(X.^2).*(Y.^nYv);
Zcc = Zcc + cf(5)*(X.^1).*(Y.^nYv);
Zcc = Zcc + cf(6)*(X.^0).*(Y.^nYv);


nYv = 0;
Zcc = Zcc + cf(7)*(X.^2).*(Y.^nYv);
Zcc = Zcc + cf(8)*(X.^1).*(Y.^nYv);
Zcc = Zcc + cf(9)*(X.^0).*(Y.^nYv);


    % Determine the estimated waviness and R^2 for each subdivided surface
Zr = Zcc;
meanZ = mean(mean(Z));
SStot = sum(sum((Z-meanZ).^2));
SSerr = sum(sum((Z-Zr).^2));
Rsq = 1 - (SSerr/SStot);


    % Subtract the estimated surface from the lesion surface to obtain
    % deviation surface
err = abs(Z-Zr);


    % Determine the surface roughness (Sa)
    Ra = mean(mean(err));


    Ra_2nd(iSd) = Ra;
    Rsq_2nd(iSd) = Rsq;
Zr2(iSd).d = Zr;


    % Sa of a subdivided surface is accepted if 0.9<= R^2 <=1.0
if and(Rsq_2nd(iSd)>=0.90,Rsq_2nd(iSd)<=1.0)
noOfAcceptedFitting = noOfAcceptedFitting + 1;
RaAcc = RaAcc + Ra_2nd(iSd);
RsqAcc = RsqAcc + Rsq_2nd(iSd);
end;
end;
% Determine the final Sa of the lesion surface
Ra_2nd_Final = NaN;
Rsq_2nd_Final = NaN;

ifnoOfAcceptedFitting>0
    Ra_2nd_Final = RaAcc/noOfAcceptedFitting;
    Rsq_2nd_Final = RsqAcc/noOfAcceptedFitting;
end;
ResMat(1,1) = Ra_2nd_Final;
ResMat(1,2) = Rsq_2nd_Final;
% ========================================================================
% ========================================================================


% ========================================================================
% ========== APPLY 3rd ORDER SURFACE FITTING TO EXTRACT ROUGHNESS =========
% ========================================================================
```

```
noOfAcceptedFitting = 0;
RaAcc = 0;
RsqAcc = 0;
foriSd=1:(noDiv^2)
    [M,N] = size(I2sDiv(iSd).Z);

    % Copy the data of subdivided surfaces into X, Y, and Z matrices
    X = I2sDiv(iSd).X(1:M,1:N);
    Y = I2sDiv(iSd).Y(1:M,1:N);
    Z = I2sDiv(iSd).Z(1:M,1:N);

    % Create matrices required for determining matrix coefficients
rowIdx = 0;
for i=1:M %noDivSp
for j=1:N %noDivSp
rowIdx = rowIdx + 1;
            Xv = X(i,j);
Yv = Y(i,j);
Zv = Z(i,j);


nYv = 3;
V(rowIdx,1)  = (Xv^3)*(Yv^nYv);
V(rowIdx,2)  = (Xv^2)*(Yv^nYv);
V(rowIdx,3)  = (Xv^1)*(Yv^nYv);
V(rowIdx,4)  = (Xv^0)*(Yv^nYv);


nYv = 2;
V(rowIdx,5)  = (Xv^3)*(Yv^nYv);
V(rowIdx,6)  = (Xv^2)*(Yv^nYv);
V(rowIdx,7)  = (Xv^1)*(Yv^nYv);
V(rowIdx,8)  = (Xv^0)*(Yv^nYv);


nYv = 1;
V(rowIdx,9)  = (Xv^3)*(Yv^nYv);
V(rowIdx,10) = (Xv^2)*(Yv^nYv);
V(rowIdx,11) = (Xv^1)*(Yv^nYv);
V(rowIdx,12) = (Xv^0)*(Yv^nYv);


nYv = 0;
V(rowIdx,13) = (Xv^3)*(Yv^nYv);
V(rowIdx,14) = (Xv^2)*(Yv^nYv);
V(rowIdx,15) = (Xv^1)*(Yv^nYv);
V(rowIdx,16) = (Xv^0)*(Yv^nYv);


Q(rowIdx,1) = Zv;
end;
end;


    % Find matrix coefficients by applying matrix inversion
VtV= V'*V;
cf = pinv(VtV)*V'*Q;


    % Create estimated waviness based on calculated matrix coefficients
Zcc = 0;
nYv = 3;
Zcc = Zcc + cf(1)*(X.^3).*(Y.^nYv);
Zcc = Zcc + cf(2)*(X.^2).*(Y.^nYv);
```

```
Zcc = Zcc + cf(3)*(X.^1).*(Y.^nYv);
Zcc = Zcc + cf(4)*(X.^0).*(Y.^nYv);
nYv = 2;
Zcc = Zcc + cf(5)*(X.^3).*(Y.^nYv);
Zcc = Zcc + cf(6)*(X.^2).*(Y.^nYv);
Zcc = Zcc + cf(7)*(X.^1).*(Y.^nYv);
Zcc = Zcc + cf(8)*(X.^0).*(Y.^nYv);


nYv = 1;
Zcc = Zcc + cf(9)*(X.^3).*(Y.^nYv);
Zcc = Zcc + cf(10)*(X.^2).*(Y.^nYv);
Zcc = Zcc + cf(11)*(X.^1).*(Y.^nYv);
Zcc = Zcc + cf(12)*(X.^0).*(Y.^nYv);


nYv = 0;
Zcc = Zcc + cf(13)*(X.^3).*(Y.^nYv);
Zcc = Zcc + cf(14)*(X.^2).*(Y.^nYv);
Zcc = Zcc + cf(15)*(X.^1).*(Y.^nYv);
Zcc = Zcc + cf(16)*(X.^0).*(Y.^nYv);


    % Determine the estimated waviness and R^2 for each subdivided surface
Zr = Zcc;
meanZ = mean(mean(Z));
SStot = sum(sum((Z-meanZ).^2));
SSerr = sum(sum((Z-Zr).^2));
Rsq = 1 - (SSerr/SStot);


    % Subtract the estimated surface from the lesion surface to obtain
    % deviation surface


err = abs(Z-Zr);
    % Determine the surface roughness (Sa)
    Ra = mean(mean(err));


    Ra_3rd(iSd) = Ra;
    Rsq_3rd(iSd) = Rsq;
Zr3(iSd).d = Zr;
    % Sa of a subdivided surface is accepted if 0.9<= R^2 <=1.0
if and(Rsq_3rd(iSd)>=0.90,Rsq_3rd(iSd)<=1.0)
noOfAcceptedFitting = noOfAcceptedFitting + 1;
RaAcc = RaAcc + Ra_3rd(iSd);
RsqAcc = RsqAcc + Rsq_3rd(iSd);
end;
end;


% Determine the final Sa of the lesion surface
Ra_3rd_Final = NaN;
Rsq_3rd_Final = NaN;


ifnoOfAcceptedFitting>0
    Ra_3rd_Final = RaAcc/noOfAcceptedFitting;
    Rsq_3rd_Final = RsqAcc/noOfAcceptedFitting;
end;


ResMat(2,1) = Ra_3rd_Final;
ResMat(2,2) = Rsq_3rd_Final;
% =========================================================================
```

```
% ========================================================================
%
% ========================================================================
% ==== SELECT THE BEST Sa BASED ON THE HIGHEST R^2 OF SURFACE FITTING =====
% ========================================================================
RsqMax = max(ResMat(:,2));
idxRa = find(ResMat(:,2)==RsqMax);
RaFinal = ResMat(idxRa(1),1);
% ========================================================================
% ========================================================================
%
% ========================================================================
% ======= DISPLAY THE FINAL RESULTS OF ROUGHNESS DETERMINATION ===========
% ========================================================================
clc;
disp(['----+-------+--------+--------+-------------+----------+--------+| ']);
disp(['PO | sbDv | Sa(mm) | Rsq | Sa sbDv(mm) | RsqsbDv |finalSa(mm)| finalRsq|']);
disp(['----+-------+--------+------+------+------+------+---------+--------| ']);

for i=1:9
polyOrd = ' ';
sRaFinal = ' ';
sRsqFinal = ' ';
    sRaFinal2n3 = ' ';
    sRsqFinal2n3 = ' ';
if i==1
polyOrd = '2nd';
sRaFinal = sprintf('%.4f',Ra_2nd_Final);
sRsqFinal = sprintf('%.4f',Rsq_2nd_Final);
        sRaFinal2n3 = sprintf('%.4f',RaFinal);
        sRsqFinal2n3 = sprintf('%.4f',RsqMax);
end;
if i==6
polyOrd = '3rd';
sRaFinal = sprintf('%.4f',Ra_3rd_Final);
sRsqFinal = sprintf('%.4f',Rsq_3rd_Final);
end;

if (i<=4)
sRa = sprintf('%.4f',Ra_2nd(i));
sRsq = sprintf('%.4f',Rsq_2nd(i));
sbDvId = num2str(i);
end;

if and((i>5),(i<=9))
        i2 = i-5;
sRa = sprintf('%.4f',Ra_3rd(i2));
sRsq = sprintf('%.4f',Rsq_3rd(i2));
sbDvId = num2str(i-5);
end;

if i ~= 5
disp([polyOrd,' | ',sbDvId,' | ',sRa,' | ',sRsq,' | ',sRaFinal,' |
    ',sRsqFinal,' | ',sRaFinal2n3,' | ',sRsqFinal2n3,' | ']);
end;
if i == 5
disp(['----+-------+--------+--------+-------------+----------+ | | ']);
end;
```

```
end;
disp(['----+-----+------+-----+---------+--------+---------+--------| ']);
% ========================================================================
% ========================================================================
close all;

figure('Position', [50, 200, 1400, 500],'Name','Orginal and Segmented Images');

subplot(231), imshow(I2d);
set(gca,'Fontsize',10,'FontWeight','bold','FontName','Arial');
title('2D Image of Skin and Lesion');

wd = maxX - minX;
hg = maxY - minY;
rectangle('Position',[minY,minX,wd,hg],'EdgeColor','green','LineWidth',2)

subplot(234), imshow(I2d(minY:maxY,minX:maxX,:));
set(gca,'Fontsize',10,'FontWeight','bold','FontName','Arial');
title('2D Image of Segmented Lesion');

% figure
subplot(232), imshow(I.Z,[]);
set(gca,'Fontsize',10,'FontWeight','bold','FontName','Arial');
title('Elevation Map of Skin and Lesion');
rectangle('Position',[minY,minX,wd,hg],'EdgeColor','green','LineWidth',2)

subplot(235), imshow(I2.Z,[]);
set(gca,'Fontsize',10,'FontWeight','bold','FontName','Arial');
title('Elevation Map of Segmented Lesion');

% figure,
subplot(233), surf(I.X,I.Y,I.Z,'edgealpha',0.1, 'facealpha',1.0);
set(gca,'Fontsize',10,'FontWeight','bold','FontName','Arial');
xlabel('X (mm)','FontWeight','bold','FontSize',10,'FontName','Arial');
ylabel('Y (mm)','FontWeight','bold','FontSize',10,'FontName','Arial');
zlabel('Z (mm)','FontWeight','bold','FontSize',10,'FontName','Arial');
title('3D View of Skin and Lesion Surfaces');
miIX = min(min(I.X));
mxIX = max(max(I.X));
miIY = min(min(I.Y));
mxIY = max(max(I.Y));
miIZ = min(min(I.Z));
mxIZ = max(max(I.Z));
axis([miIXmxIXmiIYmxIYmiIZmxIZ]);

subplot(236), surf(I2.X,I2.Y,I2.Z,'edgealpha',0.1, 'facealpha',1.0);
set(gca,'Fontsize',10,'FontWeight','bold','FontName','Arial');
xlabel('X (mm)','FontWeight','bold','FontSize',10,'FontName','Arial');
ylabel('Y (mm)','FontWeight','bold','FontSize',10,'FontName','Arial');
zlabel('Z (mm)','FontWeight','bold','FontSize',10,'FontName','Arial');
title('3D View of Segmented Lesion Surface');
miI2X = min(min(I2.X));
mxI2X = max(max(I2.X));
miI2Y = min(min(I2.Y));
mxI2Y = max(max(I2.Y));
miI2Z = min(min(I2.Z));
mxI2Z = max(max(I2.Z));
```

```
axis([miI2X mxI2X miI2Y mxI2Y miI2Z mxI2Z]);

figure('Position', [20, 50, 1500, 750],'Name','Implementation of Surface Roughness
    Algorithm on Subdivided Surfaces');
colormap(gray);
vA = -30;
vB = 50;

foridx = 1:4
idxFig = 1 + (idx-1)*5;
Zs = I2sDiv(idx);
    subplot(4,5,idxFig), surf(I2sDiv(idx).X,I2sDiv(idx).Y,Zs.Z, 'edgealpha',0.1,
    'facealpha',1.0);
view([vA,vB]);
miX(idx) = min(min(I2sDiv(idx).X));
mxX(idx) = max(max(I2sDiv(idx).X));
miY(idx) = min(min(I2sDiv(idx).Y));
mxY(idx) = max(max(I2sDiv(idx).Y));

miZ(idx) = min(min(Zs.Z));
mxZ(idx) = max(max(Zs.Z));
axis([miX(idx) mxX(idx) miY(idx) mxY(idx) miZ(idx) mxZ(idx)]);
xlabel('X (mm)','FontWeight','bold','FontSize',10,'FontName','Arial');
ylabel('Y (mm)','FontWeight','bold','FontSize',10,'FontName','Arial');
zlabel('Z (mm)','FontWeight','bold','FontSize',10,'FontName','Arial');
set(gca,'Fontsize',8,'FontWeight','bold','FontName','Arial');
    title(['SubDiv. Surface D',num2str(idx)]);
end;

foridx = 1:4
idxFig = 2 + (idx-1)*5;
subplot(4,5,idxFig),
    [m1,n1] = size(I2sDiv(idx).X);
    [m2,n2] = size(Zr2(idx).d);
    m3 = min(m1,m2);
    n3 = min(n1,n2);
    surf(I2sDiv(idx).X(1:m3,1:n3),I2sDiv(idx).Y(1:m3,1:n3),Zr2(idx).d(1:m3,1:n3),
    'edgealpha',0.1, 'facealpha',1.0);
view([vA,vB]);
miZ(idx) = min(min(Zr2(idx).d(1:m3,1:n3)));
mxZ(idx) = max(max(Zr2(idx).d(1:m3,1:n3)));
axis([miX(idx) mxX(idx) miY(idx) mxY(idx) miZ(idx) mxZ(idx)]);
xlabel('X (mm)','FontWeight','bold','FontSize',10,'FontName','Arial');
ylabel('Y (mm)','FontWeight','bold','FontSize',10,'FontName','Arial');
zlabel('Z (mm)','FontWeight','bold','FontSize',10,'FontName','Arial');
set(gca,'Fontsize',8,'FontWeight','bold','FontName','Arial');

if Rsq _ 2nd(idx)>= Rsq _ 3rd(idx)
FinZr(idx).d = Zr2(idx).d(1:m3,1:n3);
end;
if or(Rsq _ 2nd(idx)>= 0.9,Rsq _ 2nd(idx)<=1)
textColor = 'black';
end;

if or(Rsq _ 2nd(idx)< 0.9,Rsq _ 2nd(idx)>1)
textColor = 'red';
end;
```

```
title(['Estd.Wav. 2ndOrd, D',num2str(idx),' R^2=',sprintf('%.3f', Rsq _ 2nd(idx))],'
    color',textColor);
end;


sprintf('%.5f', pi)


ZeroSurf = zeros(m3,n3);


foridx = 1:4
idxFig = 3 + (idx-1)*5;
subplot(4,5,idxFig),
Er2(idx).d = Zr2(idx).d(1:m3,1:n3) - I2sDiv(idx).Z(1:m3,1:n3);
    surf(I2sDiv(idx).X(1:m3,1:n3),I2sDiv(idx).Y(1:m3,1:n3),Er2(idx).d(1:m3,1:n3),
    'edgealpha',0.1, 'facealpha',1.0);
hold on
surf(I2sDiv(idx).X,I2sDiv(idx).Y,ZeroSurf, 'edgealpha',0.1, 'facealpha',0.2,'Fac
    eColor','blue');
view([vA,vB]);
miZ(idx) = min(min(Er2(idx).d));
mxZ(idx) = max(max(Er2(idx).d));
axis([miX(idx) mxX(idx) miY(idx) mxY(idx) miZ(idx) mxZ(idx)]);
xlabel('X (mm)','FontWeight','bold','FontSize',10,'FontName','Arial');
ylabel('Y (mm)','FontWeight','bold','FontSize',10,'FontName','Arial');
zlabel('Z (mm)','FontWeight','bold','FontSize',10,'FontName','Arial');
set(gca,'Fontsize',8,'FontWeight','bold','FontName','Arial');


if Rsq _ 2nd(idx)>= Rsq _ 3rd(idx)
FinEr(idx).d = Er2(idx).d(1:m3,1:n3);
end;
if or(Rsq _ 2nd(idx)>= 0.9,Rsq _ 2nd(idx)<=1);
textColor = 'black';
end;
if or(Rsq _ 2nd(idx)< 0.9,Rsq _ 2nd(idx)>1);
textColor = 'red';
end;


title(['Dev.Surf. 2ndOrd, D',num2str(idx),' Sa =',sprintf('%.3f', Ra _ 2nd(idx)),'
    mm'],'color',textColor);
end;


foridx = 1:4
idxFig = 4 + (idx-1)*5;
subplot(4,5,idxFig),
    [m1,n1] = size(I2sDiv(idx).X);
    [m2,n2] = size(Zr3(idx).d);
    m3 = min(m1,m2);
    n3 = min(n1,n2);
    surf(I2sDiv(idx).X(1:m3,1:n3),I2sDiv(idx).Y(1:m3,1:n3),Zr3(idx).d(1:m3,1:n3),
    'edgealpha',0.1, 'facealpha',1.0);
view([vA,vB]);
miZ(idx) = min(min(Zr3(idx).d(1:m3,1:n3)));
mxZ(idx) = max(max(Zr3(idx).d(1:m3,1:n3)));
axis([miX(idx) mxX(idx) miY(idx) mxY(idx) miZ(idx) mxZ(idx)]);
xlabel('X (mm)','FontWeight','bold','FontSize',10,'FontName','Arial');
ylabel('Y (mm)','FontWeight','bold','FontSize',10,'FontName','Arial');
zlabel('Z (mm)','FontWeight','bold','FontSize',10,'FontName','Arial');
set(gca,'Fontsize',8,'FontWeight','bold','FontName','Arial');
```

```
if Rsq _ 2nd(idx)< Rsq _ 3rd(idx)
FinZr(idx).d = Zr3(idx).d(1:m3,1:n3);
end;
if or(Rsq _ 3rd(idx)>= 0.9, Rsq _ 3rd(idx)<=1);
textColor = 'black';
end;
if or(Rsq _ 3rd(idx)< 0.9, Rsq _ 3rd(idx)>1);
textColor = 'red';
end;
title(['Estd.Wav. 3rdOrd, D',num2str(idx),' R^2=',sprintf('%.3f', Rsq _ 3rd(idx))],'
    color',textColor);
end;

foridx = 1:4
idxFig = 5 + (idx-1)*5;
subplot(4,5,idxFig),
Er3(idx).d = Zr3(idx).d(1:m3,1:n3) - I2sDiv(idx).Z(1:m3,1:n3);
    surf(I2sDiv(idx).X(1:m3,1:n3),I2sDiv(idx).Y(1:m3,1:n3),Er3(idx).d(1:m3,1:n3),
    'edgealpha',0.1, 'facealpha',1.0);
hold on
surf(I2sDiv(idx).X,I2sDiv(idx).Y,ZeroSurf, 'edgealpha',0.1, 'facealpha',0.2,'Fac
    eColor','blue');
view([vA,vB]);
miZ(idx) = min(min(Er3(idx).d));
mxZ(idx) = max(max(Er3(idx).d));
axis([miX(idx) mxX(idx) miY(idx) mxY(idx) miZ(idx) mxZ(idx)]);
xlabel('X (mm)','FontWeight','bold','FontSize',10,'FontName','Arial');
ylabel('Y (mm)','FontWeight','bold','FontSize',10,'FontName','Arial');
zlabel('Z (mm)','FontWeight','bold','FontSize',10,'FontName','Arial');
set(gca,'Fontsize',8,'FontWeight','bold','FontName','Arial');

if Rsq _ 2nd(idx)< Rsq _ 3rd(idx)
FinEr(idx).d = Er3(idx).d(1:m3,1:n3);
end;
if or(Rsq _ 3rd(idx)>= 0.9, Rsq _ 3rd(idx)<=1);
textColor = 'black';
end;
if or(Rsq _ 3rd(idx)< 0.9, Rsq _ 3rd(idx)>1);
textColor = 'red';
end;

title(['Dev.Surf. 3rdOrd, D',num2str(idx),' Sa =',sprintf('%.3f', Ra _ 3rd(idx)),'
    mm'],'color',textColor);
end;

finZR =[ FinZr(1).d FinZr(2).d; FinZr(3).d FinZr(4).d];
finER =[ FinEr(1).d FinEr(2).d; FinEr(3).d FinEr(4).d];

[m4,n4] = size(finZR);
[m5,n5] = size(finER);
m6 = min(m4,m5);
n6 = min(n4,n5);

miXF = min(miX);
mxXF = max(mxX);
miYF = min(miY);
```

```
mxYF = max(mxY);
miZF = min(miZ);
mxZF = max(mxZ);

figure('Position', [50, 200, 1400, 500],'Name','Final Result of Surface Roughness
    Algorithm');
colormap(gray)
subplot(131), surf(I2.X(1:m6,1:n6),I2.Y(1:m6,1:n6),I2.Z(1:m6,1:n6),'edgealpha',0.1,
    'facealpha',1.0);
view([vA,vB]);
miXF = min(min(I2.X(1:m6,1:n6)));
mxXF = max(max(I2.X(1:m6,1:n6)));
miYF = min(min(I2.Y(1:m6,1:n6)));
mxYF = max(max(I2.Y(1:m6,1:n6)));
miZF = min(min(I2.Z(1:m6,1:n6)));
mxZF = max(max(I2.Z(1:m6,1:n6)));
axis([miXFmxXFmiYFmxYFmiZFmxZF]);
xlabel('X (mm)','FontWeight','bold','FontSize',12,'FontName','Arial');
ylabel('Y (mm)','FontWeight','bold','FontSize',12,'FontName','Arial');
zlabel('Z (mm)','FontWeight','bold','FontSize',12,'FontName','Arial');
set(gca,'Fontsize',10,'FontWeight','bold','FontName','Arial');
title('Lesion Surface');

subplot(132), surf(I2.X(1:m6,1:n6),I2.Y(1:m6,1:n6),finZR(1:m6,1:n6),'edgealpha',0.1,
    'facealpha',1.0);
view([vA,vB]);
axis([miXFmxXFmiYFmxYFmiZFmxZF]);
xlabel('X (mm)','FontWeight','bold','FontSize',12,'FontName','Arial');
ylabel('Y (mm)','FontWeight','bold','FontSize',12,'FontName','Arial');
zlabel('Z (mm)','FontWeight','bold','FontSize',12,'FontName','Arial');
set(gca,'Fontsize',10,'FontWeight','bold','FontName','Arial');
title(['Estimated Waviness, R^2 =',sprintf('%.3f', RsqMax)]);

ZeroSurf = zeros(m6,n6);
subplot(133), surf(I2.X(1:m6,1:n6),I2.Y(1:m6,1:n6),finER(1:m6,1:n6),'edgealpha',0.1,
    'facealpha',1.0);
hold on
surf(I2.X(1:m6,1:n6),I2.Y(1:m6,1:n6),ZeroSurf, 'edgealpha',0.1, 'facealpha',0.2,'F
    aceColor','blue');
view([vA,vB]);
miZF = 2*min(min(finER));
mxZF = 2*max(max(finER));
axis([miXFmxXFmiYFmxYFmiZFmxZF]);
xlabel('X (mm)','FontWeight','bold','FontSize',12,'FontName','Arial');
ylabel('Y (mm)','FontWeight','bold','FontSize',12,'FontName','Arial');
zlabel('Z (mm)','FontWeight','bold','FontSize',12,'FontName','Arial');
set(gca,'Fontsize',10,'FontWeight','bold','FontName','Arial');
title(['Deviation Surface, Sa =',sprintf('%.3f', RaFinal),' mm']);
```

References

1. Pivarcsi A, Nagy I, Kemeny L. Innate immunity in the skin: how keratinocytes
   fight against pathogens. *Current Immunology Reviews* 2005;1:29–42.
2. Faller A, Schunke M, Schunke G, Taub E. *The human body: an introduction to structure and function.* New York: Georg Thieme, 2004.

3. American Association for the Advancement of Science. *The science inside skin.* Washington, DC: American Association for the Advancement of Science, 2004.
4. Lewis R. *Life.* Boston: McGraw-Hill, 2004.
5. Society for Dermopharmacy. The life cycle of a horny cell. http://www.scf-online.com/english/35_e/frontpage35_e.htm
6. Lutz Slomianka. Blue histology — integumentary system. http://www.lab.anhb.uwa.edu.au/mb140/corepages/integumentary/integum.htm
7. Overney GT. Human histology for amateur microscopists. http://www.micros-copy-uk.org.uk/mag/artaug02/gohisto.html
8. Griffiths CEM, Barker JNWN. Pathogenesis and clinical features of psoriasis. *Lancet* 2007;370:263–271.
9. Lowes MA, Bowcock AM, Krueger JG. Pathogenesis and therapy of psoriasis. *Nature* 2007;445:866–873.
10. McCracke GA, Eilers D. Psoriasis. *Medical Update for Psychiatrists* 1997;2:78–80.
11. Bhalerao J, Bowcock AM. The genetics of psoriasis: a complex disorder of the skin and immune system. *Human Molecular Genetics* 1998;7:1537–1545.
12. Goodless D. Symptoms of psoriasis. http://psoriasis.about.com/od/symp-tomsdiagnosis/tp/symptomsofpsoriasis.htm
13. Zygote Media Group. Solid 3D male body model. http://www.3dscience.com/3D_Models/Human_Anatomy/Solid_Models/solid-3d-male-model
14. LEO Pharma. Where can psoriasis be located? http://www.psorinfo.com/Locations.aspx?ID=80
15. Langley RGB, Krueger GG, Griffiths CEM. Psoriasis: epidemiology, clinical fea-tures, and quality of life. *Annals of the Rheumatic Diseases* 2005;64:ii18–23; discus-sion ii24–25.
16. Weinberg JM. Psoriasis. In: Hall BJ, Hall JC, eds. *Sauer's manual of skin diseases.* Riverwoods, IL: Lippincott Williams & Wilkins, 2010:160–163.
17. Feldman SR, Krueger GG. Psoriasis assessment tools in clinical trials. *Annals of the Rheumatic Diseases* 2005;64:65–68.
18. Bonsard V, Paul C, Prey S, Puzenat E, Gourraud P-A, Aractingi S, Aubin F, Bagot M, Cribier B, Joly P, Jullien D, Le Maitre M, Richard-Lallemand M-A, Ortonne J-P. What are the best outcome measures for assessing plaque psoriasis sever-ity? A systematic review of the literature. *Journal of the European Academy of Dermatology and Venereology* 2010;24(Suppl 2):17–22.
19. Fredriksson T, Pettersson U. Severe psoriasis—oral therapy with a new retinoid. *Dermatologica* 1978;157:238–244.
20. Buettner PG, Garbe C. Agreement between self-assessment of melanocytic nevi by patients and dermatologic examination. *American Journal of Epidemiology* 2000;151:72–77.
21. Masters M, McMahon M, Svens B. Reliability testing of a new scar assessment tool, Matching Assessment of Scars and Photographs (MAPS). *Journal of Burn Care Rehabilitation* 2005;26:273–284.
22. Better Medicine. Skin symptoms. http://www.localhealth.com/article/skin-symptoms
23. RightDiagnosis. Causes of skin lesions. http://www.rightdiagnosis.com/symptoms/skin_lesion/causes.htm
24. MedicineNet. Definition of lesion. http://www.medterms.com/script/main/art.asp?articlekey=4135

25. MacNeal RJ. Description of skin lesions. http://msd-bahamas.com/mmpe/sec10/ch109/ch109b.html

26. GFMesstechnik. *User manual: PRIMOS optical 3D skin measuring device*. Berlin: GFMesstechnik, 2008.

27. Hani A, Prakasa E, Fitriyah H, Nugroho H, Affandi A, Hussein S. High order polynomial surface fitting for measuring roughness of psoriasis lesion. In: Badioze Zaman H, Robinson P, Petrou M, Olivier P, Shih T, Velastin S, Nyström I, eds. *Visual informatics: sustaining research and innovations*. Berlin: Springer, 2011:341–351.

28. Dong WP, Mainsah E, Stout KJ. Reference planes for the assessment of surface roughness in three dimensions. *International Journal of Machine Tools and Manufacture* 1995;35:263–271.

29. Hani AF, Prakasa E, Nugroho H, Affandi AM, Hussein S. Sample area for surface roughness determination of skin surfaces. 4th International Conference on Intelligent and Advanced Systems (ICIAS), vol. 1. Universiti Teknologi PETRONAS, Kuala Lumpur, 2012:328–332.

30. Yoon JS, Ryu C, Lee JH. Developable polynomial surface approximation to smooth surfaces for fabrication parameters of a large curved shell plate by differential evolution. *Computer-Aided Design* 2008;40:905–915.

31. Borradaile GJ. *Statistics of earth science data: their distribution in time, space, and orientation*. Berlin: Springer, 2003.

32. Barbato Carneiro K, Garnaes J, Gori G, Hughes G, Jensen CP, Jørgensen JF, Jusko O, Livi S, McQuoid H, Nielsen L, Picotto GB, Wilkening GG. *Scanning tunnelling microscopy methods for roughness and micro hardness measurements. Synthesis report for the European Union under its Programme for Applied Metrology*. CDNA-16145. Brussels: European Commission.

33. Hani M, Fadzil A, Prakasa E. Methodology and apparatus for objective, non-invasive and in vivo assessment and rating of psoriasis lesion scaliness using digital imaging. U.S. Patent 20,120,308,096.

34. Ahmad Fadzil MH, Prakasa E, Fitriyah H, Nugroho H, Affandi A, Hussein S. Validation on 3D surface roughness algorithm for measuring roughness of psoriasis lesion. *International Jounral of Biological and Life Sciences* 2012;8:205–210.

35. Sizes.com. Grades of sandpaper (coated abrasives). http://www.sizes.com/tools/sandpaper.htm

36. McNaught AD, Wilkinson A. *IUPAC compendium of chemical terminology*. Oxford: Blackwell Scientific, 1997.

37. Bland J, Altman D. Statistical methods for assessing agreement between two methods of clinical measurement. *Lancet* 1986;1:307–310.

38. Jain AK, Murty MN, Flynn PJ. Data clustering: a review. *ACM Computing Surveys* 1999;31:264–323.

39. Jain AK. Data clustering: 50 years beyond K-means. *Pattern Recognition Letters* 2010;31:651–666.

40. Leung S-H, Wang S-L, Lau W-H. Lip image segmentation using fuzzy clustering incorporating an elliptic shape function. *IEEE Transactions on Image Processing* 2004;13:51–62.

41. Chahir Y, Elmoataz A. Skin-color detection using fuzzy clustering. *Proceedings of the ISCCSP* 2006;3:1–4.

42. Zhou H, Schaefer G, Sadka AH, Celebi ME. Anisotropic mean shift based fuzzy c-means segmentation of dermoscopy images. *IEEE Journal of Selected Topics in Signal Processing* 2009;3:26–34.
43. Cucchiara R, Grana C. Using the Topological Tree for skin lesion structure description. Sixth International Conference on Knowledge-based Intelligent Information and Engineering Systems (KES 2002), vol. 1, Crema, Italy, September 16–18, 2002.
44. Bezdek JC, Ehrlich R, Full W. FCM: the fuzzy c-means clustering algorithm. *Computers & Geosciences* 1984;10:191–203.
45. Zadeh LA. Fuzzy sets. *Information and Control* 1965;8:338–353.
46. Xu R, Wunsch DC. Clustering algorithms in biomedical research: a review. *IEEE Reviews in Biomedical Engineering* 2010;3:120–154.
47. Zanaty EA. Determining the number of clusters for kernelized fuzzy c-means algorithms for automatic medical image segmentation. *Egyptian Informatics Journal* 2012;13:39–58.
48. Landis, JR and GG Koch, "The Measurement of Observer Agreement for Categorical Data," *Biometrics* 1977; vol. 33, pp 159–174.

2

Determination of Lesion Color for Clustering Psoriasis Erythema

Ahmad Fadzil Mohamad Hani and Esa Prakasa

CONTENTS

2.1 Introduction on Psoriasis Erythema

Psoriasis is an incurable but treatable skin disease and can exist for a long time [1]. The disease is characterized by an abnormally high growth rate of new skin cells. Under normal conditions, skin cells will grow in approximately twenty-eight days, but in psoriasis the growth can occur in four days [2]. Psoriasis is marked by the appearance of red lesions on body surfaces. At greater severities the lesion thickness increases, with coarse, slivery scales appearing. Psoriasis is not a contagious disease. Recent studies have shown that psoriasis can significantly affect quality of life, with many psoriasis patients experiencing social and psychological problems [3]. The published psoriasis prevalence in Malaysia by the Dermatological Society of Malaysia is 3% [4]. Of the 75,883 patients registered in the Hospital Kuala Lumpur from 2005 to 2010, 3,906 psoriasis patients were registered in the

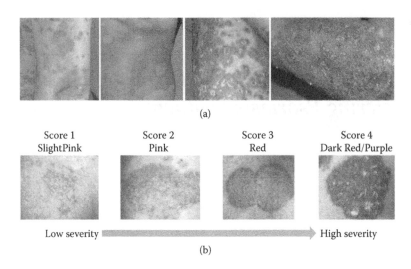

FIGURE 2.1
(a) Red plaques of psoriasis lesions. (b) Erythema severity scores.

Dermatology Department, resulting in an incidence of 5.2% [5]. Several examples of red plaques of psoriasis lesions and erythema severity levels are depicted in Figure 2.1.

There are several types of psoriasis, including guttate, pustular, flexural, nail, erythrodermic, and psoriatic arthritis [6]. The Malaysian Psoriasis Registry found that plaque psoriasis is the most common type in Malaysia, at 80% [7]. Psoriasis can affect the skin surfaces in any body regions. However, there are some regions that have a higher probability to be affected by psoriasis, such as the elbows, knees, scalp, and lower back [8]. Periodic reviews of medical treatments for psoriatic patients is important, as the disease cannot be completely cured. The Psoriasis Area and Severity Index (PASI) scoring method, the gold standard for psoriasis assessment, is applied to grade psoriasis severities [9]. The total PASI score is determined based on four assessment parameters. These parameters are lesion area, erythema, scaliness, and thickness. These parameters are assessed separately on four body regions: head, upper limbs, trunk, and lower limbs.

In daily practice, dermatologists determine scores of PASI parameters based on visual and tactile perceptions. The scores are then weighted by ratio coefficients and added up to find the total PASI score. The total PASI score ranges from 0 to 72. A 75% reduction in PASI score is required for a treatment to be declared effective [9]. Equation (2.1) is used to compute the PASI score.

$$
\begin{aligned}
\text{PASI} = {}& 0.1 \times (E_h + T_h + S_h)A_h + 0.2 \times (E_u + T_u + S_u)A_u \\
&+ 0.3 \times (E_t + T_t + S_t)A_t + 0.4 \times (E_l + T_l + S_l)A_l
\end{aligned}
\tag{2.1}
$$

The score ranges of PASI parameters are area $(A) = 0$–6, erythema $(E) = 0$–4, thickness $(T) = 0$–4, and scaliness $(S) = 0$–4. The subscripts h, u, t, and l are used to represent head, upper limbs, trunk, and lower limbs, respectively.

2.2 Erythema Measurement

Although the PASI scoring method has been considered the gold standard for psoriasis assessment, it is not used in daily clinical practice. PASI scoring is tedious and time-consuming, and can be very subjective. The four PASI parameters must be determined for four body regions, which means a total of sixteen assessments. The scores given by dermatologists are subjective and influenced by inter- and intrarater variation. To minimize these variations, an objective assessment method is required that is reliable, valid, and consistent from assessment to assessment.

Skin erythema can be measured in several ways. In one of the methods, erythema is measured based on the amount of microvascular red blood cells under the skin using a laser Doppler flowmeter (LDF). In another method, erythema is measured based on the lesion's absorbance and reflectance with either broadband or selected wavelengths in the visible range. Erythema can also be measured by analyzing the tristimulus values of light reflected from skin structures [7, 8].

Erythema is affected by the number of blood cells under the skin. The LDF is the first erythema meter used in dermatological applications. This equipment measures capillary blood perfusion parameters, such as blood flow, velocity, and volume. To measure these parameters, a small area of tissue that contains red blood and stationary tissue cells is illuminated by laser light. The light is randomly scattered and reflected by both cell types. The scattered and reflected lights are received by a detector, as shown in Figure 2.2. The moving blood cells in the capillaries cause a Doppler shift in the incident light, resulting in fluctuations in the detector signal. The power spectrum of the detector signal is then interpreted to measure proportional flow, velocity, and volume of the blood cells [9].

Hemoglobin and melanin are the most dominant pigments affecting the color of the skin. Hemoglobin is located in the blood capillaries of the dermis, whereas melanin is found in the keratinocytes and melanocytes of the epidermis. Hemoglobin absorbs more light at shorter wavelengths, with a peak in the green wavelengths, and absorbs very little at the longer wavelengths. Hence blood is visually perceived as a red color. Melanin absorbs light of all wavelengths, increasingly at shorter wavelengths, without a peak at any particular wavelength. As erythema increases, a greater amount of green light is absorbed, and thus less light is reflected. Based on these differences of melanin and hemoglobin in the absorbance spectrum, Diffey et al. [10] proposed

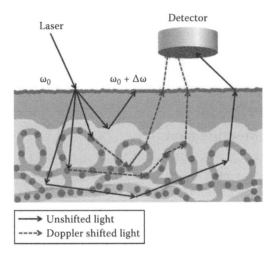

FIGURE 2.2
Principle of laser Doppler flowmeter.

a method to estimate the melanin and erythema indexes using the following equations:

$$M = \log_{10}\left(\frac{1}{\% \text{ red reflectance}} \right) \tag{2.2}$$

$$E = \log_{10}\left(\frac{1}{\% \text{ green reflectance}} \right) - \log_{10}\left(\frac{1}{\% \text{ red reflectance}} \right) \tag{2.3}$$

The derma spectrometer is a narrowband reflectometer that is used to determine the melanin (M) and erythema (E) indexes. Red and green light-emitting diodes (LEDs) are used to measure the reflectance spectrum. The red light is centered at 655 nm, with a half-width of 30 nm, and the green light is centered at 568 nm, with a half-width of 30 nm. The measured skin area is illuminated with the two light sources sequentially. An example of a derma spectrometer is shown in Figure 2.3.

The International Commission on Illumination (CIE), an international authority on light, illumination, color, and color spaces, has modeled the three components—light source, reflectance characteristic of the object, and color vision of the human eye—that affect the overall color of an object. The CIE created the CIELAB color space to describe color perceived by humans. Color is described by three stimulus values: L^*, a^*, and b^*. L^* represents brightness, varying from 0 to 100; a^* represents the degree of greenness–redness (negative values indicate green and positive values red); and b^* represents the degree of blueness–yellowness (negative values indicate blue and positive

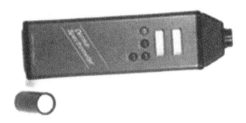

FIGURE 2.3
The derma spectrometer.

values yellow). A colorimeter instrument such as a chromameter can be used to measure color in terms of the CIELAB tristimuli. The chromameter consists of a xenon lamp that emits white light as the light source and has three high-sensitivity silicone photocells acting as the detector of reflected light. The sensitivity of the photocells is adjusted to match human eye sensitivity.

The aforementioned modalities (LDF, spectrometer, and chromameter) have been used in dermatology to assess erythema. In one study, two types of chromameter—Lange Micro Color and Minolta Chromameter CR-200—and an LDF were used to quantify erythema involving eighteen volunteers, ages eighteen to seventy-two years old. The measurement results were compared with clinician's scoring [11]. Erythema was elicited by a local irritant, sodium lauryl sulfate.

A total of fifty-five patch tests of erythema and eighteen patch tests of normal skin were assessed. The instruments assessed erythema and normal skin patches individually and also the differences between the two patches. Clinical scores were obtained by comparing the erythema patch with a normal skin patch from the same body region. The scores range from 0 to 4, with 0 indicating no erythema and 4 indicating marked erythema through out the test area. The results showed that with erythema there is a clear and highly significant increase in a values, some decrease in L^* values, and a significant increase in LDF values. Clinical scoring correlated positively with the increase in a^* values of both chromameters and an increase in LDF. However, LDF is sensitive to physical activity of the patient during measurement, such as talking. Thus measurement consistency is harder to achieve with this instrument compared with the measurement by chromameter. This phenomena has been also observed by Lahti et al. [7].

Both a^* and E have been used by dermatologists to represent skin erythema [10–12]. However, Shriver and Parra [13] found that the relationship between a^* and E is complex and dependent on the level of pigmentation. A clear positive correlation between a^* and E is found only in persons with low melanin content ($M < 40$). In order to understand this phenomenon, correlation between the E and M indexes and between a^* and L^* are analyzed. A negative correlation between the E and M indexes is observed in patients with high pigmentation levels ($M > 40$) as shown in Figure 2.4(a). There is

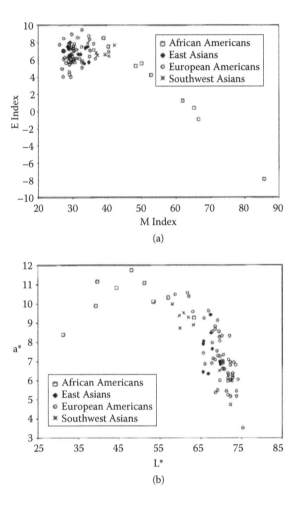

FIGURE 2.4
Relationship between (a) the E and M indexes and between (b) L^* and a^* for the inner upper arm regions [13].

no correlation found in patients with low-pigmented skin ($M < 40$). This indicates that E is independent from M, thus it becomes a good indicator of erythema in low-pigmented skin. On the other hand, a negative correlation between a^* and L^* is found in patients with low pigmentation levels ($L^* > 60$) as shown in Figure 2.4(b). Thus a^* is a good indicator of erythema for high-pigmented skin.

Draaijers et al. [14] evaluated the reliability of visual assessment, chromameter, and derma spectrometer in assessing the color of scars. Four observers assessed forty-nine scar areas on twenty different patients independently. Each observer rated the difference in pigmentation and vascularization of the scar compared with normal skin according to the Patient

 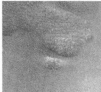  

| Fair skin with light red/pink lesion | Light brown skin with red lesions | Brown skin with darker red lesions | Dark skin with dark purplish lesions |

FIGURE 2.5
Erythema of various skin tones.

and Observer Scar Assessment Scale (POSAS). The scores range from 1 to 10; the higher the score, the worse the scar. In addition, observers also categorized the pigmentation type of the scar into normal, hypopigmentation, mixed pigmentation, or hyperpigmentation, represented by scores of 0 to 3, respectively. Spearman's ρ correlation coefficient was calculated between the visual assessment and instrument measurements. Correlations between the visual assessment of vascularization and E and a^* were found to be significant but weak ($r = 0.5$ and 0.42, respectively). Correlations between the visual assessment of pigmentation type and M, L^*, and b^* were found to be low ($r = 0.32$, 0.23, and 0.24, respectively).

The measurement reliabilities of the derma spectrometer, chromameter and visual assessment were analyzed by means of the intraclass correlation coefficient (ICC) and its 95% confidence interval [14]. Although the quality of the measurements of the instruments appeared comparable or better than the visual assessment, the analysis for erythema scoring still needs to be developed.

In summary, the current methods of measuring erythema suffer from problems such as inconsistency of measurements due to human activity when using the LDF, low correlations between visual assessments and erythema index calculations, and the need for different indicators for skin tones of low- and high-pigmented skin. Figure 2.5 shows the different skin tones and the erythema colors of psoriasis lesions. Patients are grouped into fair, light brown, brown, and dark skins. Depending on the skin tone of the normal skin and the degree of lesion severity, psoriasis lesions appear in varying pink to red and purplish colors. These problems require the use of measuring instruments and analysis approaches that are independent of activity and the skin tone of the subject to measure erythema effectively.

A non-invasive computerized system for monitoring and grading PASI erythema has been developed in order to overcome the above issues. The system uses a chromameter for stable and consistent skin color data measurement. The chromameter can be connected to a computer system that performs the analysis and classification for determining the PASI score. Figure 2.6(a) shows a Konica chromameter CR-400 that is used in assessing

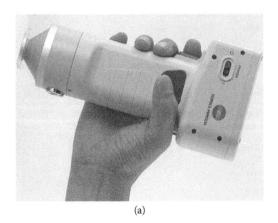

(a)

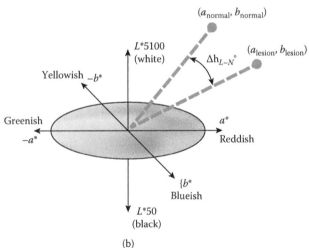

(b)

FIGURE 2.6
(a) A Konica chromameter CR-400. (b) The CIELAB color space and hue differences.

lesion erythema. The CIELAB color space is used to determine lesion color for in vivo psoriasis erythema scoring. The CIELAB is defined as a color space that is perceptually linear with the human visual system.

The hue difference (Δh_{L-N}) between a lesion sample and its corresponding normal skin sample is measured as follows: First, the hue of surrounding normal skin (h_N) and the lesion sample (h_L) are calculated using equations (2.4) and (2.5).

$$h_N = \tan^{-1}(b_N^*/a_N^*)$$ (2.4)

$$h_L = \tan^{-1}(b_L^*/a_L^*)$$ (2.5)

The hue difference, Δh_{L-N}, is then defined as in equation (2.6):

$$\Delta h_{L-N} = |h_L - h_N| \qquad (2.6)$$

By taking the difference between the hues of the normal skin and the psoriasis lesion as the erythema value, the measured erythema will be independent of the hue of the normal skin. This will overcome the problem of having different indicators for different skin tones.

2.3 Development of the PASI Erythema Classification

2.3.1 Fuzzy c-Means (FCM) Clustering for Erythema Scoring

In data management, the clustering algorithm is required since the volume and the variety of data are increasing rapidly nowadays. The clustering algorithm helps the user to understand, process, and summarize this abundant data [15]. Clustering problems happen in various applications. The problems are not limited in engineering areas only, but can be found in the economic, social, biologic, and medical fields. The clustering algorithm is required to provide a decision based on the processed data. The decision might be used either in the intermediate stages or in the final stage of the system. By observing the clustered data, the analysis can be carried out more easily and in a more organized manner.

As mentioned by Jain et al. [16], the clustering is considered unsupervised classification. For supervised classification it is referred to as discriminant analysis. The difference between these approaches is in the initial condition of the training data. For the supervised classification, the training data are labeled and used to define the cluster descriptions. These descriptions are then used to classify the inputted data. In contrast, the unsupervised classification or clustering uses unlabeled data for the training. The data are clustered based on similarity criteria. The process is performed iteratively until the centroids are maximally separated [16].

Unsupervised clustering algorithms have been widely applied for image analyses of medical applications. Several applications are mentioned in the following section. Li and Yuen [17] apply a regularized color clustering algorithm to find significant colors for tongue color diagnosis. A total of 6,788 red, green, and blue (RGB) color blocks were extracted from the tongue images of sixty-four patients. The blocks were clustered to obtain twenty-five significant color groups. By applying the algorithm, the intercluster distance can be increased, whereas the intracluster distance can be minimized. These color groups are used to classify eight medical categories [17].

Masulli and Schenone [18] developed the possibilistic neuro fuzzy c-means algorithm (PNFCM) to segment brain features of magnetic resonance imaging (MRI) images. This algorithm was developed by combing the possibilistic c-means (PCM-II) and the capture effect neural network (CENN). The CENN is added to estimate the number of clusters automatically. Fuzzy clustering has advantages in dealing with the segmentation problems at tissue boundaries. The PNFCM has been evaluated by applying the algorithm to two datasets. The clustering results are compared with manual segmentation made by a group of skilled clinicians. For the first evaluation, the algorithm is able to segment the brain structures of normal people. In the second evaluation, the algorithm is able to locate tumor areas on brain images of meningioma patients. The PNFCM has shown high accuracies in the first and second evaluations. The average accuracy of these evaluations is 83.88%. The white matter, gray matter, eyes, skull, edema, and tumor areas are successfully segmented [18].

Clustering algorithms are widely applied to recognize and classify data into a number of clusters. The k-means and the FCM clustering algorithms are widely applied in clustering implementation. FCM is principally an improved form of the k-means clustering algorithm.

The term k-means was proposed by James MacQueen in 1967, based on the idea of a Polish mathematician, Hugo Steinhaus, in 1956. The algorithm was then standardized by Stuart Lloyd in 1957, but it was not published until 1982. The algorithm was implemented for pulse-code modulation. Meanwhile, E. W. Forgy also proposed a similar clustering method in 1965. Therefore the k-means algorithm is also known as the Lloyd–Forgy method. Although the k-means algorithm was introduced more than fifty years ago, the algorithm is still widely applied for data clustering. This algorithm is simple, efficient, and easy to implement [15]. The k-means algorithm groups the dataset into a predefined number of clusters. Initial centroids are assigned randomly or by applying certain algorithms to the clusters. Similarities of the dataset and the centroids are examined for every single data point and the data are assigned to a cluster that has the highest similarity. After all of the datasets are clustered, new centroids are recalculated by finding the mean values of the clusters. The process is iterated until the centroids are considered constant. This consistency is indicated by minimum changes at all the centroids. A certain number of iterations can also be used to define the termination point of the iteration process.

The main difference between k-means and FCM is in the probability of the data belonging to a certain cluster. In k-means, data belong to a cluster only. Thus there are only two options for its probability, probability equals one at its cluster and zero for the other clusters. In contrast to k-means, the FCM algorithm defines the data probabilities for each defined cluster. These probabilities are known as membership degrees. The function that used to determine the degrees is referred to as a membership function. The FCM defines different membership degrees for each data probability to the existing clusters. The degrees might be calculated by applying the input data to the available membership functions.

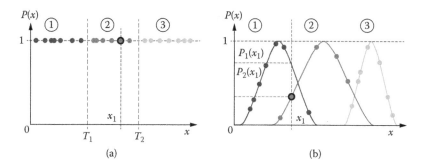

FIGURE 2.7
Comparison of classification methods used in (a) hard and (b) soft clustering algorithms.

Figure 2.7 shows the principle difference in the classification methods is hard clustering (k-means) and soft clustering (FCM). For the hard clustering, an input, x_1, is classified as cluster 2 since its value is located inside of two boundary levels (from T_1 to T_2). There are only two options for the membership degree, either 0 or 1. Figure 2.7(a) shows that the membership degree of x_1 belonging to cluster 2 is 1, whereas the membership degrees of x_1 for the other clusters is 0. In soft clustering, as shown by Figure 2.7(b), an input x_1 can be considered as a member of three clusters (clusters 1, 2, and 3). However, it has different membership degrees. For the example, the membership degrees of x_1 belonging to clusters 1, 2, and 3 are 0.8, 0.4, and 0.0, respectively. The classification system decides x_1 as a member of cluster 1. The decision is taken since the membership degree at cluster 1 is the highest among the clusters.

A k-means classification system is built based on the final centroids obtained from the clustering process. The system classifies input data based on the distance between the input data and the cluster centroid. The data belong to a cluster if the data have the highest similarity to the cluster, which is represented by the cluster centroid. FCM clustering was initially proposed by Bezdek et al. [19]. The algorithm was developed based on the fuzzy set proposed by Zadeh [20]. Bezdek et al. introduced this algorithm to improve the k-means algorithm. FCM minimizes the limitations of hard clustering algorithms, such as k-means clustering, by associating each piece of data to all existing clusters with a degree of membership [21]. The membership degree is represented by a membership matrix $U = [u_{ik}]$. The membership degree of a data point in a particular cluster can be considered as a probability that this data point belongs to the cluster. Subscripts i and k are used to index the data point and the cluster, respectively. For instance, if there are a hundred data points ($i = 1, ..., 100$) and three clusters ($k = 1, 2, 3$), then the matrix size of U becomes 100×3. The membership degrees can be any real values, but they have to be between 0 and 1. A data point will be associated with each cluster and it is determined by a membership degree. The accumulation of these degrees is always equal to 1. The constraints of the membership degree are described in equation (2.7) [22].

$$\sum_{k=1}^{K} u_{ik} = 1, \quad \forall i = 1, \ldots, N; 0 \le u_{ik} \le 1 \tag{2.7}$$

The iteration process of the FCM algorithm is performed to achieve membership degree stability. This condition indicates the data points have been clustered and maximally separated. The FCM algorithm uses an objective function to control the iteration process. The objective function of the FCM algorithm is formulated as written in equation (2.8). This equation is used to classify a dataset by applying only a single feature parameter. A new variable, the fuzziness coefficient (m), is introduced to the equation. The aim of this coefficient is to symbolize the fuzziness of the membership function. This fuzziness represents the softness of the cluster boundaries. As a comparison, in hard clustering the fuzziness coefficient is zero. It denotes the firmness of the cluster boundaries for hard clustering. The variable K is the number of clusters, whereas N is the number of data points. Centroid initialization can be executed either randomly or based on prior knowledge.

$$J(C,U,X) = \sum_{k=1}^{K} \sum_{i=1}^{N} u_{ik}^{m} |x_i - c_k| \tag{2.8}$$

The iteration process updates the centroids and the membership matrix continuously. The equation for calculating the new centroids is given by equation (2.9).

$$c_k = \frac{\sum_{i=1}^{N} u_{ik}^{m} x_i}{\sum_{i=1}^{N} u_{ik}^{m}} \tag{2.9}$$

The membership matrix is determined by applying equation (2.10).

$$u_{ik} = \frac{1}{\sum_{l=1}^{K} \left[\frac{|x_i - c_k|}{|x_i - c_l|} \right]^{2/(m-1)}}$$

$$i = 1, 2, \ldots, N \quad k = 1, 2, \ldots, K \tag{2.10}$$

The objective function is going to be stable when high membership degrees are assigned near to the cluster centroids, while low membership degrees are obtained if the data are far from the centroids [22]. The iteration process to update matrixes U and M is repeated until there are no changes in the cluster centroids. It is indicated by $\Delta J < \varepsilon$, where ε is the acceptable difference to terminate the iteration. The membership degree of a data point in a particular cluster can be considered as a probability that this data point belongs to the cluster.

Algorithm 2.1: The FCM algorithm

1. Define the number of clusters (K), membership matrix ($u_{i,k}$), and the fuzziness coefficient (m). Use small random numbers for the membership matrix u_{ik} and a real value between 1 and 2 for the fuzziness coefficient. Fix the objective function $J(C)$ equal to 0. The acceptable error (ε) and the maximum number of iterations are also initialized in this step. These values are used to indicate the termination point of the iteration.

2. Compute new centroids by applying u_{ik} and x_i to equation (2.9).

3. Update the new membership matrix based on new centroids in step 2 and equation (2.10).

4. Determine the objective function $J(C)$ by applying the centroids and the membership matrix to equation (2.8).

5. Determine the difference between the latest and the previous objective function. The difference is denoted as ΔJ. If the difference is larger than the predefined acceptable error (ε), then continue the iteration to step 3. The iteration could be terminated if $\Delta J < \varepsilon$ or the iteration number is greater than the maximum value.

6. End.

Figure 2.8 displays a flowchart of the FCM algorithm. The number of clusters (K), fuzziness coefficient (m), acceptable error (ε), and initial objective function $J(C)$ are considered as clustering and iteration parameters. The fuzziness coefficient is used to shape the fuzziness level of the clusters. It is suggested that the fuzziness coefficient be set greater than 1. If the coefficient is equal to 1, then there is no fuzziness in the clusters. It implies the cluster separation is crisp and the cluster does not overlap with the other clusters. By applying a higher coefficient ($m > 1$), the cluster boundaries become softer and enable cluster overlapping. The fuzziness coefficient usually is set to 2, which is an optimum value to provide fuzzy clusters with separable capability [23]. From the clustering result, it is found that the membership degrees are not calculated at every data point value outside the values given in the training stage. Therefore a membership function is required to determine membership degrees at any data point. The membership degrees of the cluster are used to create a membership function. Curve fitting algorithms can be applied to find the best membership function. Gaussian functions might be used to fit the membership degrees of the clustered data points.

Figure 2.9 describes the iteration process of a training dataset that is clustered into three groups. A set of random small numbers is used to initiate the membership degrees, as show in Figure 2.9. Configurations of membership

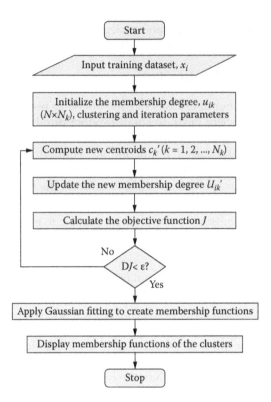

FIGURE 2.8
Flowchart of the FCM algorithm.

degrees at the middle (4th) and final (15th) iteration steps are depicted in Figure 2.9(b and c), respectively. Spline lines are used in Figure 2.9(a–c) to show the grouping of membership degrees. In the final stage, Gaussian functions are applied to fit these clustered membership degrees. Finally, in classifier implementation, input data are classified into a particular cluster, which has the highest membership degree. Here the FCM algorithm is applied to determine PASI erythema scores [24, 25].

2.3.2 Skin Tone Classification

Figure 2.10 displays a lesion sample that is considered as a representative lesion for erythema assessment. The normal skin sample is obtained from the skin surrounding the lesion. Black circles indicate the areas where erythema measurements need to be performed. The chromameter is used to obtain the L^*, a^*, and b^* values of the lesion and normal skin areas.

The skin tones of normal skin can be categorized into dark, brown, light brown, and fair skin tones. Skin tones are represented by the L^* values of measured skin color data using the chromameter. The FCM algorithm, an unsupervised

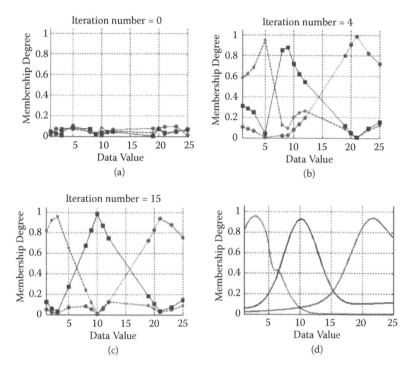

FIGURE 2.9
The changes in membership degrees at the (a) initial, (b) middle, (c) final iteration steps, and (d) Membership functions of clustered membership degrees.

clustering approach, is applied to divide the dataset into four clusters. The dataset consists of 1,462 normal skin samples that are used to build a skin tone classifier. Membership degrees of the training dataset can be obtained from clustering results. In the final iteration, each data point will have four membership degrees that represent the degree of belonging to a particular cluster. Figure 2.11 shows

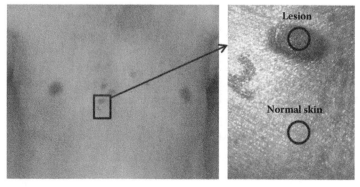

FIGURE 2.10
Lesion and surrounding normal skin.

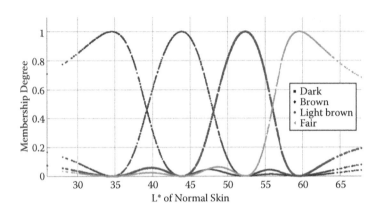

FIGURE 2.11
Membership degrees of the skin tone groups.

clustered membership degrees of skin tone groups. There are some discontinuity points in all the clusters. This condition occurs because the skin tone data at these points are not available in the training dataset. Membership functions are required to resolve these discontinuity problems.

Since the membership degrees are still discrete, general equations are required to determine skin tone at any L^* values. The membership degrees shown in Figure 2.12 are then fitted by applying Gaussian fitting functions. The membership functions of skin tone groups are shown in Figure 2.12. The coefficient of determination (R^2) is used to indicate the best fit of the Gaussian functions to the clustered dataset.

The general Gaussian equation is given by equation (2.11) and the coefficients for four skin tone groups are given in Table 2.1. The equation is built by three Gaussian functions.

$$T_n(L) = a_1\exp\left(-\left(\frac{L-b_1}{c_1}\right)^2\right) + a_2\exp\left(-\left(\frac{L-b_2}{c_2}\right)^2\right) + a_3\exp\left(-\left(\frac{L-b_3}{c_3}\right)^2\right) \quad (2.11)$$

To classify skin tone with an L^* value, skin tone probabilities of the skin tone groups have to be determined. An input L^* value is classified as a member of a skin tone group if this group can provide the highest probability among the skin tone groups. The rule of skin tone classification can be written as follows:

$$T(L) = \begin{cases} 1, \ \max(T_1(L), T_2(L), T_3(L), T_4(L)) = T_1(L) \\ 2, \ \max(T_1(L), T_2(L), T_3(L), T_4(L)) = T_2(L) \\ 3, \ \max(T_1(L), T_2(L), T_3(L), T_4(L)) = T_3(L) \\ 4, \ \max(T_1(L), T_2(L), T_3(L), T_4(L)) = T_4(L) \end{cases} \quad (2.12)$$

TABLE 2.1

Coefficients of Membership Functions for Skin Tone Classification

Skin Tone Group	a_1	b_1	c_1	a_2	b_2	c_2	a_3	b_3	c_3	R^2
Dark (T_1)	0.529	34.563	3.727	0.769	28.289	7.352	0.355	37.708	2.778	0.998
Brown (T_2)	0.867	44.119	3.293	0.439	40.452	2.810	0.278	47.250	2.443	0.995
Light brown (T_3)	0.809	52.648	2.751	0.516	49.455	2.649	0.313	55.214	2.078	0.993
Fair (T_4)	0.545	59.857	3.087	0.760	65.248	6.146	0.394	57.231	2.238	0.997

2.3.3 Erythema Score Classification

For the development of the PASI erythema classification, the dataset is divided in two: the training and testing datasets. The training dataset comprises of lesion color data in which the measurements of the color data are taken only once. There are 1,462 lesions in the training dataset. The test dataset contains color data of lesions that are obtained by measuring the lesions twice. There are 430 lesions in the test dataset, resulting in 860 color data. The objective of the double measurement is to evaluate the performance of erythema assessment. The training dataset is used to obtain the membership functions of the erythema scores. The membership functions are approximated by using the Gaussian equation, which is composed of two Gaussian functions, as shown in equation (2.13). Table 2.2 lists the coefficients of the Gaussian functions for erythema scoring for dark, brown, light brown, and fair skin tones.

$$S_n(\Delta h) = a_1 \exp\left(-\left(\frac{\Delta h - b_1}{c_1}\right)^2\right) + a_2 \exp\left(-\left(\frac{\Delta h - b_2}{c_2}\right)^2\right) \tag{2.13}$$

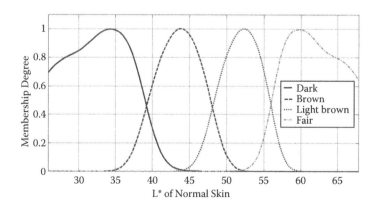

FIGURE 2.12
Membership functions of the skin tone groups.

TABLE 2.2

Coefficients of Gaussian Functions for Erythema Classification of Dark, Brown, Light Brown, and Fair Skin Tones ($N_{Total} = 1462$)

Skin Tone	Erythema Score	a_1	b_1	c_1	a_2	b_2	c_2	R^2
	1	0.718	2.183	2.238	3.372×10^{12}	−194.214	35.717	0.991
Dark	2	0.097	6.856	0.595	0.897	6.645	2.625	0.976
$n = 261$	3	0.454	10.648	1.842	0.596	12.292	3.428	0.972
	4	0.781	17.848	3.342	4.648×10^{12}	309.295	52.498	0.994
	1	0.715	3.709	3.333	3.391×10^{12}	−281.321	51.826	0.991
Brown	2	0.155	10.972	1.718	0.825	10.182	3.936	0.971
($n = 360$)	3	0.476	16.161	2.581	0.581	18.539	5.136	0.975
	4	0.795	26.751	5.005	3.765×10^{12}	446.057	75.749	0.992
	1	0.694	4.838	3.714	2.320×10^{12}	−361.156	67.078	0.989
Light brown	2	0.097	12.774	0.978	0.898	12.531	4.548	0.972
($n = 487$)	3	0.161	20.046	1.795	0.829	20.860	5.065	0.977
	4	0.657	29.276	4.277	5.879×10^{-1}	38.422	11.350	0.987
	1	0.756	6.967	4.977	2.574×10^{12}	−407.796	75.567	0.992
Fair	2	0.371	17.735	3.087	0.640	15.865	5.686	0.976
($n = 405$)	3	0.088	24.761	0.987	0.905	25.052	4.580	0.969
	4	0.663	32.930	3.906	6.865×10^{-1}	43.661	11.962	0.986

For each skin tone there are four membership functions of erythema scores provided. Figures 2.13–2.16 depict the clustered dataset and membership functions of erythema scores for dark, brown, light brown, and fair skin tones.

To score erythema with a given hue difference value, hue difference probabilities of the skin tone groups have to be determined. An input hue difference value is classified as a member of the erythema score group if this group can provide the highest probability among the erythema score groups. The rule of erythema scoring can be written as follows:

$$S(\Delta h) = \begin{cases} 1, \max(S_1(\Delta h), S_2(\Delta h), S_3(\Delta h), S_4(\Delta h)) = S_1(\Delta h) \\ 2, \max(S_1(\Delta h), S_2(\Delta h), S_3(\Delta h), S_4(\Delta h)) = S_2(\Delta h) \\ 3, \max(S_1(\Delta h), S_2(\Delta h), S_3(\Delta h), S_4(\Delta h)) = S_3(\Delta h) \\ 4, \max(S_1(\Delta h), S_2(\Delta h), S_3(\Delta h), S_4(\Delta h)) = S_4(\Delta h) \end{cases} \quad (2.14)$$

The complete process for the developed PASI erythema assessment that comprises the skin tone classification (fair, light brown, brown, and dark) and erythema classification stages (scores 1–4) is depicted in Figure 2.17.

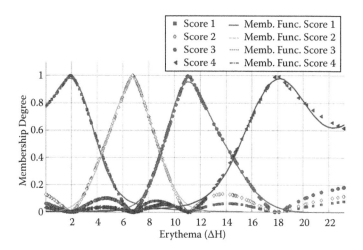

FIGURE 2.13
Fitted membership functions of the PASI erythema score for dark skin tone.

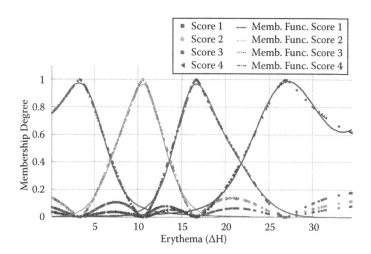

FIGURE 2.14
Fitted membership functions of the PASI erythema score for brown skin tone.

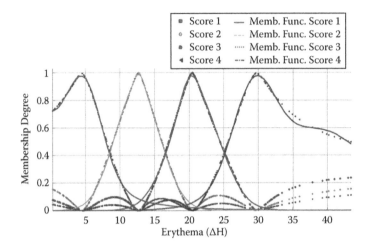

FIGURE 2.15
Fitted membership functions of the PASI erythema score for light brown skin tone.

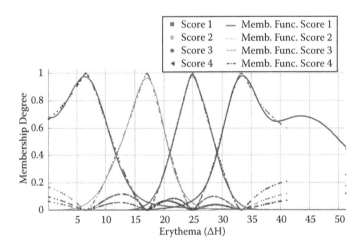

FIGURE 2.16
Fitted membership functions of the PASI erythema score for fair skin tone.

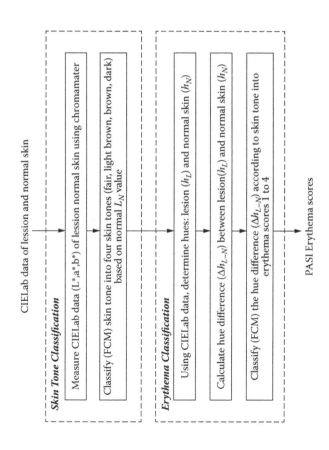

FIGURE 2.17
The PASI erythema assessment.

2.4 Agreement Analysis of PASI Erythema

To evaluate the performance of the developed PASI erythema assessment, agreement analysis between two assessments by the developed method is compared to the agreement analysis of assessments by two dermatologists. The kappa coefficient values of the PASI erythema scores on forty-eight patients were determined for different skin tones. Several representative lesions are initially marked by the dermatologist for each patient. In erythema assessment, one region is represented by one lesion. Therefore, in one session the dermatologist performs four erythema assessments of four body regions. To conduct assessments in two sessions, erythema assessment on all body regions (head, upper limbs, trunk, lower limbs) needs to be completed in the first assessment. The same procedures are then repeated for the second session. Figure 2.18 describes a session of PASI erythema assessment. Four PASI erythema scores are obtained from this assessment.

Scoring agreement among the raters is analyzed in this section. Table 2.3 shows that the kappa coefficients of the PASI erythema computerized system are higher than those of the dermatologists. Moreover, the kappa coefficients are greater than 0.75 for light brown, brown, and dark skin tones. Only for the fair skin tone is the kappa coefficient below 0.75 (0.7). This is due to the fact that the fair skin tone is sensitive.

As seen in Figure 2.19, there are many acquisition error cases involving fair skin tone in comparison with other skin tones (acquisition error of fair skin tone is symbolized by brown circles). The acquisition error occurred because the exact area measured by the chromameter is not known. Measurement procedure can only approximate the target area since the measured area is covered by the chromameter sensor. This is worsened by the fact that the lesion is very small. Figure 2.19 shows the plot of the first versus second measurements. Ideally data points of the first and second measurements

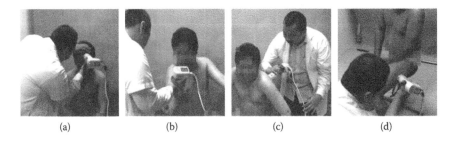

| (a) | (b) | (c) | (d) |

FIGURE 2.18
A session of PASI erythema assessment consists of measurement on (a) head, (b) upper limb, (c) trunk, and (d) lower limb.

TABLE 2.3

Kappa Coefficients for Different Skin Tones

Patient	Skin Tone	Dermatologist 1 Versus Dermatologist 2		First Assessment Versus Second Assessment	
		N (lesion)	kappa	N (lesion)	kappa
1	Fair	64	0.66	64	0.70
2	Light brown	103	0.60	103	0.80
3	Brown	48	0.69	48	0.79
4	Dark	28	0.70	28	0.90
Total N/average kappa	243		0.66	243	0.82

should be on the reference line ($Y = X$). To ensure the precision of the measurements is high, only data within 1 standard deviation (σ) of the reference line are used in the analysis. The reference lines for ±1 standard deviation (σ) are illustrated in Figure 2.19.

The overall kappa coefficient (0.82) is high and shows that the PASI erythema system has a high reliability. With this high reliability, the system can be potentially used for monitoring erythema severities, particularly for light brown, brown, and dark skin tones. To minimize the acquisition error, several successive measurements must be taken during the data acquisition and only data in the interval of 1 standard deviation (σ) from the reference line ($Y = X$ line) are used. This requirement can be introduced in the erythema color data acquisition protocol.

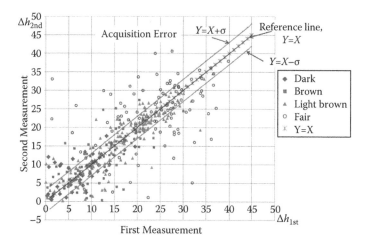

FIGURE 2.19

Acquisition error of the first and second measurements.

2.5 Conclusion

A PASI erythema assessment system has been developed. The system uses a chromameter for skin data measurement that performs the analysis and classification for psoriasis erythema. The method comprises two stages: skin tone classification of normal skin followed by erythema assessment of the lesion. The two stages thus provide a more robust assessment of erythema, which vary according to the skin tone of the subject.

It is shown that dermatologist agreement (kappa coefficient) is found to be less than 0.71 for erythema. Dermatologist assessments for PASI erythema scores are therefore not considered reliable and as a result, the designs of classifiers for PASI scores are based on unsupervised methods. The FCM clustering method has been proposed as a classification system for both skin tone and erythema. For PASI erythema, the kappa coefficients for erythema assessment are greater than 0.7 for all skin tones (fair, 0.7; light brown, 0.8; brown, 0.79; dark skin, 0.9). This shows that the method has an almost perfect agreement for erythema assessment of dark skin tone and substantial agreement for erythema assessment of fair, light brown, and brown skin tones. From this performance evaluation, it can be concluded that the developed method is reliable for clinical use.

Appendix: Matlab Code

FCM Training Stage

```
% = = = = = = = = = = = LOAD THE TRAINING DATASET = = = = == = = = = = = = = =
clc;clear all; close all;
[filename, pathname, filterindex] = uigetfile('train*.mat', 'Pick an M-file');
load([pathname,filename]);
clustOpt = 1;%%% (1) For skin tone clustering
            %%%(2) For erythema clustering of fair skin tone
switchclustOpt
case 1
        x = L;
Nk = 4;% number of cluster, Dark, Brown, Light Brown, Fair
case 2
        x = dH4;
Nk = 4;% number of cluster, Score 1, 2, 3, and 4
end;


% = = = = =  INITIALISE MEMBERSHIP DEGREES AND ITERATION PARAMETERS = = = = = = =
%- - - - - - - - - - - - - - - - - - - - - - - - - - - - - - - - - - - - - - -
Nk = 4;
N = length(x);
% membership degrees initialisation using random numbers
u0 = rand(N,Nk)/10;
u = u0;
```

```
% fuzziness coefficient
% Choose fuzziness coefficient. Best values are between 1.4 and 2.8 from:
% http://fuzziness.org/fcm- > Check again with another references!
m = 3.5;
OldJm = 1e9;
deltaJm = 1e9;
noI = 0;
gaussOrd = 3;% 1,2, or 3 Gaussian functions will be used to create a membership
             % function
%- - - - - - - - - - - - - - - - - - - - - - - - - - - - - - - - - - - - - - -
% = = =   = = CLUSTERING ITERATION = = =  = = = = = = = = = = = = = = = = = = =
%- - - - - - - - - - - - - - - - - - - - - - - - - - - - - - - - - - - - - - -
% while deltaJm> 1e-5
whiledeltaJm> 1e-5
noI = noI + 1;

%%%- - - DETERMINE THE CLUSTER CENTROIDS- - - - - - - - - - - - - - - -
    % centroids determination
for j = 1:Nk
          A = 0;
          B = 0;
for i = 1:N
                A = A + (u(i,j)^m*x(i));
                B = B + u(i,j)^m;
end;
C(j) = A/B;
end;
%%%- - - - - - - - - - - - - - - - - - - - - - - - - - - - - - - - - - - - - -

%%%- - - COMPUTE & UPDATE MEMBERSHIP DEGREES OF TRAINING DATASET- - - - -
pw = 2/(m-1);
for j = 1:Nk
for i = 1:N
              A = 0;
for l = 1:Nk
                  A0 = abs(x(i)-C(j));
                  B0 = abs(x(i) - C(l));
                  A = A + (A0/B0)^pw;
end;
u(i,j) = 1/A;
end;
end;
%%%- - - - - - - - - - - - - - - - - - - - - - - - - - - - - - - - - - - - - -

%%%- - - CALCULATE THE OBJECTIVE FUNCTION- - - - - - - - - - - - - - - -
Jm = 0;
for i = 1:N
for j = 1:Nk
Jm = Jm + (u(i,j)^m)*(x(i)-C(j))^2;
end;
end;

    % Update on objective function
deltaJm = abs(Jm - OldJm);
OldJm = Jm;

    % Show the latest objective function in the command window
JmA(noI) = Jm;
disp([num2str(noI),' : ',num2str(Jm)]) ;
```

```
for i = 1:Nk
c(i).data = ones(N,1)*C(i);
end;


JmFinalMat = ones(N,1)*Jm;
finalMat = [];
finalMat = [x u];
for i = 1:Nk
finalMat = [finalMat c(i).data];
end;
finalMat = [finalMatJmFinalMat];


finalMatStruct(noI).data = finalMat;
%%%- - - - - - - - - - - - - - - - - - - - - - - - - - - - - - - - - - -
end;


% = = = = = FIT THE MEMBERSHIP DEGREES OF CLUSTERED DATASET = = = = = = =
% = = = = == = = = = = = INTO GAUSSIAN FUNCTION = = = == = = = = = = = =
%- - - - - - - - - - - - - - - - - - - - - - - - - - - - - - - - - - - -
noMaxI = length(finalMatStruct);
finalMatData = finalMatStruct(noMaxI).data;
finalMatData = sortrows(finalMatData,1);
xData = finalMatData(:,1);


for i = 1:Nk
idxMat(i,1) = i+1;
idxMat(i,2) = idxMat(i,1)+Nk;
idxMat(i,3) = finalMatData(1, (Nk+1+i));
end;
idxMat2 = sortrows(idxMat,3);


for i = 1:Nk
idxMatInt = floor(idxMat2(i));
mFunc(i).d = finalMatData(:,idxMatInt);
end;
centroids = sort(C);


x = xData;
miX = min(xData); mxX = max(xData);
% GAUSSIAN SUMS (Peak fitting)
% MODELNAME      EQUATION
% gauss2         Y = a1*exp(-((x-b1)/c1)^2)+a2*exp(-((x-b2)/c2)^2)


for i = 1:Nk
    y = mFunc(i).d;
switchgaussOrd
case 1
gaussType = 'gauss1';
case 2
gaussType = 'gauss2';
case 3
gaussType = 'gauss3';
end;


    [cfun, rsquare] = fit(x,y,gaussType);
cfunCls(i).data = cfun;
rsquareCls(i).data = rsquare;
```

```
end;
%- - - - - - - - - - - - - - - - - - - - - - - - - - - - - - - - - - - -


% = = = = = = DISPLAY THE FIGURES OF CLUSTERING RESULTS = = = = = = = = =
%- - - - - - - - - - - - - - - - - - - - - - - - - - - - - - - - - - - -
figure('Position',[50, 50, 1000, 900],'Name','FCM Clustering Results');
subplot(211);
colTyp(1) = 'g';
colTyp(2) = 'b';
colTyp(3) = 'r';
colTyp(4) = 'm';
colTyp(5) = 'k';
colTyp(6) = 'c';


%%%- - - MEMBERSHIP DEGREES OF CLUSTERED DATASET- - - - - - - - - - - -
subplot(211);
for i = 1:Nk
if i = =1
plot(xData, mFunc(i).d,'LineStyle','none','Marker','square',...
        'MarkerEdgeColor',[0 164 0]/255,'MarkerFaceColor',[0 164 0]/255,...
        'MarkerSize',3);
end;
if i~ = 1
plot(xData, mFunc(i).d,'LineStyle','none','Marker','square',...
        'MarkerEdgeColor',colTyp(i),'MarkerFaceColor',colTyp(i),...
        'MarkerSize',3);
end;
hold all
end;
grid on
axis([miXmxX 0 1.2]);
labelFontSize = 12;
xlabel('Data Value','FontSize',labelFontSize,'fontweight','b')
ylabel('Membership degree','FontSize',labelFontSize,'fontweight','b')
set(gca,'Fontsize',11,'FontWeight','bold','FontName','Arial');
title('Membership degree of clustered dataset','FontSize',...
labelFontSize,'fontweight','b')
%- - - - - - - - - - - - - - - - - - - - - - - - - - - - - - - - - - - -


%%%- - - MEMBERSHIP FUNCTIONS OF CLUSTERED DATASET- - - - - - - - - - - -
subplot(212);
xC = linspace(miX,mxX,100);
for i = 1:Nk
fitR_sq = rsquareCls(i).data.rsquare;
iffitR_sq> = 0.9
switchgaussOrd
case 1
            a1 = cfunCls(i).data.a1; b1 = cfunCls(i).data.b1;
            c1 = cfunCls(i).data.c1;
fittedMemFunc = a1*exp(-((xC-b1)/c1).^2);
case 2
            a1 = cfunCls(i).data.a1; b1 = cfunCls(i).data.b1;
            c1 = cfunCls(i).data.c1;
            a2 = cfunCls(i).data.a2; b2 = cfunCls(i).data.b2;
            c2 = cfunCls(i).data.c2;
fittedMemFunc = a1*exp(-((xC-b1)/c1).^2) + a2*exp(-((xC-b2)/c2).^2);
case 3
            a1 = cfunCls(i).data.a1; b1 = cfunCls(i).data.b1;
            c1 = cfunCls(i).data.c1;
```

```
                a2 = cfunCls(i).data.a2; b2 = cfunCls(i).data.b2;
                c2 = cfunCls(i).data.c2;
                a3 = cfunCls(i).data.a3; b3 = cfunCls(i).data.b3;
                c3 = cfunCls(i).data.c3;
fittedMemFunc = a1*exp(-((xC-b1)/c1).^2) + a2*exp(-((xC-b2)/c2).^2)...
                + a3*exp(-((xC-b3)/c3).^2);
end;
end;
if i = =1
plot(xC,fittedMemFunc,'LineWidth',2,'Color',[0 164 0]/255),
end;
if i>1
plot(xC,fittedMemFunc,'LineWidth',2,'Color',colTyp(i)),
end;


hold all
end;


plotTitle = '';
for i = 1:Nk
tVal = ['C(',num2str(i),') = ',num2str(centroids(i)),' '];
plotTitle = [plotTitletVal];
end;
title(plotTitle);
xlabel('Data Value','FontSize',labelFontSize,'fontweight','b')
ylabel('Membership degree','FontSize',labelFontSize,'fontweight','b')
set(gca,'Fontsize',11,'FontWeight','bold','FontName','Arial');
title('Membership function of clustered dataset','FontSize',...
labelFontSize,'fontweight','b')
grid on
axis([miXmxX 0 1.2]);
%- - - - - - - - - - - - - - - - - - - - - - - - - - - - - - - - - - -


savemembFunctCoef.mat x gaussOrdNkrsquareClscfunCls centroids colTyp;
```


FCM Classification

```
clear all; clc;
clc;
% = = = = = = = CLASSIFY AN INPUT DATA = = = = = = = = = = = = = = = = = =
%- - - - - - - - - - - - - - - - - - - - - - - - - - - - - - - - - - - -
%18 PASI00002 LL
xM(1,1:8) = [38.640  8.960   13.430  46.400  3.650   4.860   3.198   1];
%499    64    PASI00001    LL
xM(2,1:8) = [45.326  11.672 17.217   30.308  4.876   5.086   9.658   2];
%1084   68    PASI00003 UL
xM(3,1:8) = [53.027  9.398   19.150  52.297  14.648  13.282  21.660  3];
%       36    PASI00001 LL
xM(4,1:8) = [57.987  10.263  19.106  57.373  12.887  10.629  22.242  4];


disp('- -+- - - - - - - - - - +- - - - - - - - - - - ');
disp(['No',' ',' LNorm ',' aNorm ',' bNorm ','| LLes ',' aLes ',' bLes ']);
disp('- -+- - - - - - - - - - +- - - - - - - - - - - ');
for i = 1:4
textDisp = [num2str(i),' '];
for j = 1:6
valDisp = sprintf('%2.3f',xM(i,j));
```

```
if length(valDisp) = = 5
valDisp = [' ',valDisp];
end;
textDisp = [textDisp,' ',valDisp];
end;
disp(textDisp);
end;
disp(['5 ',' Input a new data ']);
disp('- -+- - - - - - - - - - - +- - - - - - - - - - - - ');


inputOpt = input('Please select your option, 1, 2, 3, 4, or 5? ');

if and(inputOpt> = 1,inputOpt< = 4)
L_normal = xM(inputOpt,1);
a_normal = xM(inputOpt,2);
b_normal = xM(inputOpt,3);
L_lesion = xM(inputOpt,4);
a_lesion = xM(inputOpt,5);
b_lesion = xM(inputOpt,6);
end;


ifinputOpt = =5
L_normal = input('L of normal skin? (Exmp: 43.542) ');
a_normal = input('a of normal skin? (Exmp: 11.780) ');
b_normal = input('b of normal skin? (Exmp: 18.569) ');
disp('- - - - - - - - - - - - - - - ');
    % = = = = = = = = = = = Lesion = = = = = = = = = = = = = = =
L_lesion = input('L of skin lesion? (Exmp: 46.892) ');
a_lesion = input('a of skin lesion? (Exmp: 13.980) ');
b_lesion = input('b of skin lesion? (Exmp: 18.884) ');
end;
for k = 1:4
switch k
case 1
        a1 =     -0.3028;%  (-0.375, -0.2306)
        b1 =       31.96;%  (31.77, 32.14)
        c1 =       2.335;%  (2.004, 2.667)
        a2 = 4.83e+011;%  (-1.219e+015, 1.22e+015)
        b2 =       -1507;%  (-1.329e+005, 1.299e+005)
        c2 =         285;%  (-1.172e+004, 1.229e+004)
        a3 =      0.9531;%  (0.8655, 1.041)
        b3 =       33.44;%  (33.05, 33.83)
        c3 =       5.824;%  (5.548, 6.099)
case 2
        a1 =      0.1941;%  (0.1768, 0.2115)
        b1 =       44.72;%  (44.65, 44.78)
        c1 =      0.9112;%  (0.7995, 1.023)
        a2 = 5.013e+011;%  (-2.576e+015, 2.577e+015)
        b2 =       -2995;%  (-5.343e+005, 5.283e+005)
        c2 =       560.6;%  (-4.842e+004, 4.954e+004)
        a3 =      0.6979;%  (0.6843, 0.7115)
        b3 =       44.26;%  (44.22, 44.3)
        c3 =       4.571;%  (4.474, 4.668)
case 3
        a1 =      0.1756;%  (0.1597, 0.1915)
        b1 =       52.38;%  (52.34, 52.43)
        c1 =      0.7361;%  (0.6516, 0.8207)
        a2 = 3.817e+011;%  (-2.687e+015, 2.688e+015)
        b2 =        3927;%  (-9.367e+005, 9.445e+005)
```

```
          c2 =        719.7;%  (-8.659e+004, 8.802e+004)
          a3 =       0.7039;%  (0.692, 0.7158)
          b3 =        52.29;%  (52.26, 52.32)
          c3 =        3.902;%  (3.825, 3.978)
case 4
          a1 =       0.1503 ;%  (0.1111, 0.1896)
          b1 =        59.47;%  (59.34, 59.6)
          c1 =        1.001;%  (0.7466, 1.255)
          a2 =        1.223;%  (0.9656, 1.481)
          b2 =        62.33;%  (61.73, 62.93)
          c2 =        6.095;%  (5.895, 6.295)
          a3 =      -0.5675;%  (-0.8472, -0.2877)
          b3 =        63.24;%  (62.83, 63.65)
          c3 =        3.389;%  (2.472, 4.306)
end;
T(k) = a1*exp(-((L_normal-b1)/c1).^2) + a2*exp(-((L_normal-b2)/c2).^2)+
a3*exp(-((L_normal-b3)/c3).^2);
end;


tone = find(T = =max(T));
switch tone
case 1
disp('Skin tone: Dark');
case 2
disp('Skin tone: Brown');
case 3
disp('Skin tone: Light Brown');
case 4
disp('Skin tone: Fair');
end;


% = = = = = CALCULATE HUE OF NORMAL SKIN AND SKIN LESION = = = = = = = = =
%- - - - - - - - - - - - - - - - - - - - - - - - - - - - - - - - - - - -
aV = a_normal;
bV = b_normal;
hDeg = atand(abs(bV/aV));
%% quadrant 1
if and(aV> = 0,bV> = 0)
    h = hDeg;
end;
%% quadrant 2
if and(aV<0,bV> = 0)
    h = 180 - hDeg;
end;
%% quadrant 3
if and(aV<0,bV<0)
    h = 180 + hDeg;
end;
%% quadrant 4
if and(aV> = 0,bV<0)
    h = 360 - hDeg;
end;
h_normal = h;
%- - - - - - - - - - - - - - - - - - - - - - - - - - - - - - - - - - - -
% = = = = = = = HUE OF SKIN LESION = = = = = = = = = = = = = = = = = = = =
%- - - - - - - - - - - - - - - - - - - - - - - - - - - - - - - - - - - -
aV = a_lesion;
bV = b_lesion;
```

```
hDeg = atand(abs(bV/aV));
%% quadrant 1
if and(aV> = 0,bV> = 0)
    h = hDeg;
end;
%% quadrant 2
if and(aV<0,bV> = 0)
    h = 180 - hDeg;
end;
%% quadrant 3
if and(aV<0,bV<0)
    h = 180 + hDeg;
end;
%% quadrant 4
if and(aV> = 0,bV<0)
    h = 360 - hDeg;
end;
h_lesion = h;
%- - - - - - - - - - - - - - - - - - - - - - - - - - - - - - - - - - - - -

% = = = CALCULATE HUE DIFFERENCE BETWEEN LESION AND NORMAL SKIN = = =  = =
%- - - - - - - - - - - - - - - - - - - - - - - - - - - - - - - - - - - - -
dH = abs(h_normal - h_lesion);
%- - - - - - - - - - - - - - - - - - - - - - - - - - - - - - - - - - - - -

% dH = 10;

switch tone
case 1
    %Dark
cf = [1.063 1.642 2.923;
    1.039 6.7 2.463;
    1.049 11.65 2.877;
    1.095 19.23 4.731;];

case 2
    %Brown
cf = [1.08 2.83 4.329;
    1.044 10.36 3.581;
    1.065 17.57 4.152;
    1.093 28.61 6.619;];

case 3
    %Light brown
cf = [1.074 3.772 4.987;
    1.048 12.5 4.203;
    1.033 20.660 4.564;
    1.088 31.93 7.18;];

case 4
    %Fair
cf = [1.073 5.431 6.746;
    1.043 16.54 4.768;
    1.042 25.090 4.398;
    1.115 35.05 5.959;];
end;
```

```
for i = 1:4
Sc(i) = cf(i,1)*exp(-((dH-cf(i,2))/cf(i,3)).^2);
end;
score = find(Sc = =max(Sc));
disp(['Erythema score: ',num2str(score)]);
% = = = = = DISPLAY MEMBERSHIP FUNCTIONS AND THE MEASURED TONE = = = = = =
%- - - - - - - - - - - - - - - - - - - - - - - - - - - - - - - - - - - - -
figure('Position',[50, 50, 400, 400],'Name','Skin Tone Clustering');
colTyp(1) = 'b';
colTyp(2) = 'r';
colTyp(3) = 'g';
colTyp(4) = 'k';


LArr = linspace(30,70,200);
N = length(LArr);
for k = 1:4
switch k
case 1
        a1 =    -0.3028;%  (-0.375, -0.2306)
        b1 =      31.96;%  (31.77, 32.14)
        c1 =      2.335;%  (2.004, 2.667)
        a2 = 4.83e+011;%  (-1.219e+015, 1.22e+015)
        b2 =       -1507;%  (-1.329e+005, 1.299e+005)
        c2 =        285;%  (-1.172e+004, 1.229e+004)
        a3 =     0.9531;%  (0.8655, 1.041)
        b3 =      33.44;%  (33.05, 33.83)
        c3 =      5.824;%  (5.548, 6.099)
case 2
        a1 =     0.1941;%  (0.1768, 0.2115)
        b1 =      44.72;%  (44.65, 44.78)
        c1 =     0.9112;%  (0.7995, 1.023)
        a2 = 5.013e+011;%  (-2.576e+015, 2.577e+015)
        b2 =      -2995;%  (-5.343e+005, 5.283e+005)
        c2 =      560.6;%  (-4.842e+004, 4.954e+004)
        a3 =     0.6979;%  (0.6843, 0.7115)
        b3 =      44.26;%  (44.22, 44.3)
        c3 =      4.571;%  (4.474, 4.668)
case 3
        a1 =     0.1756;%  (0.1597, 0.1915)
        b1 =      52.38;%  (52.34, 52.43)
        c1 =     0.7361;%  (0.6516, 0.8207)
        a2 = 3.817e+011;%  (-2.687e+015, 2.688e+015)
        b2 =       3927;%  (-9.367e+005, 9.445e+005)
        c2 =      719.7;%  (-8.659e+004, 8.802e+004)
        a3 =     0.7039;%  (0.692, 0.7158)
        b3 =      52.29;%  (52.26, 52.32)
        c3 =      3.902;%  (3.825, 3.978)
case 4
        a1 =     0.1503;%  (0.1111, 0.1896)
        b1 =      59.47;%  (59.34, 59.6)
        c1 =      1.001;%  (0.7466, 1.255)
        a2 =      1.223;%  (0.9656, 1.481)
        b2 =      62.33;%  (61.73, 62.93)
        c2 =      6.095;%  (5.895, 6.295)
        a3 =    -0.5675;%  (-0.8472, -0.2877)
        b3 =      63.24;%  (62.83, 63.65)
        c3 =      3.389;%  (2.472, 4.306)
end;
for i = 1:N
```

```
TArr(k,i) = a1*exp(-((LArr(i)-b1)/c1).^2) + a2*exp(-((LArr(i)-b2)/c2).^2)+
a3*exp(-((LArr(i)-b3)/c3).^2);
end;
end;
% subplot(211);
for k = 1:4
plot(LArr,TArr(k,:),'LineWidth',2,'Color',colTyp(k));
xlabel('L of normal skin','FontSize',11,'fontweight','b');
ylabel('Membership degree','FontSize',11,'fontweight','b');
set(gca,'Fontsize',11,'FontWeight','bold','FontName','Arial');
axis([min(LArr) max(LArr) 0 1.1]);

grid on;
hold on
end;
switch tone
case 1
title('Skin tone: Dark');
case 2
title('Skin tone: Brown');
case 3
title('Skin tone: Light Brown');
case 4
title('Skin tone: Fair');
end

xLn = L_normal*ones(1,10);
yLn = linspace(0,1.1,10);
line(xLn,yLn,'Color','m','LineWidth',2.5,'LineStyle','-.');

figure('Position',[500, 50, 400, 400],'Name','Erythema Clustering');

dHArr = linspace(0,43,200);
N = length(LArr);
switch tone
case 1
    %Dark
cf = [1.063 1.642 2.923;
    1.039 6.7 2.463;
    1.049 11.65 2.877;
    1.095 19.23 4.731;];

case 2
    %Brown
cf = [1.08 2.83 4.329;
    1.044 10.36 3.581;
    1.065 17.57 4.152;
    1.093 28.61 6.619;];

case 3
    %Light brown
cf = [1.074 3.772 4.987;
    1.048 12.5 4.203;
    1.033 20.660 4.564;
    1.088 31.93 7.18;];

case 4
    %Fair
```

```
cf = [1.073 5.431 6.746;
    1.043 16.54 4.768;
    1.042 25.090 4.398;
    1.115 35.05 5.959;];
end;

for k = 1:4
for i = 1:N
ScArr(k,i) = cf(k,1)*exp(-((dHArr(i)-cf(k,2))/cf(k,3)).^2);
end;
end;
% subplot(212);
for k = 1:4
plot(dHArr,ScArr(k,:),'LineWidth',2,'Color',colTyp(k));
xlabel('Erythema (dH)','FontSize',11,'fontweight','b');
ylabel('Membership degree','FontSize',11,'fontweight','b');
set(gca,'Fontsize',11,'FontWeight','bold','FontName','Arial');

grid on;
hold on
end;
switch score
case 1
title('Erythema score: 1');
case 2
title('Erythema score: 2');
case 3
title('Erythema score: 3');
case 4
title('Erythema score: 4');
end

xLn = dH*ones(1,10);
yLn = linspace(0,1.1,10);
line(xLn,yLn,'Color',[64 0 0]/255,'LineWidth',2.5,'LineStyle','-');
axis([min(dHArr) max(dHArr) 0 1.1]);
```

References

1. National Psoriasis Foundation. Psoriasis. http://www.psoriasis.org/netcommunity/ learn_statistics
2. Lippincott Williams & Wilkins. *Pathophysiology: a 2-in-1 reference for nurses.* Philadelphia: Lippincott Williams & Wilkins, 2005.
3. Bhosle MJ, Kulkarni A, Feldman SR, Balkrishnan R. Quality of life in patients with psoriasis. *Health and Quality of Life Outcomes* 2006;4:35.
4. Overview of psoriasis in Malaysia. http://www.dermatology.org.my/overview_psoriasis.html
5. Hospital Kuala Lumpur. Registered patients in Hospital Kuala Lumpur.
6. Neimann AL, Porter SB, Gelfand JM. The epidemiology of psoriasis. *Expert Review of Dermatology* 2006;1:63–76.

7. Lahti A, Kopola H, Harila A, Myllylä R, Hannuksela M. Assessment of skin erythema by eye, laser Doppler flowmeter, spectroradiometer, two-channel erythema meter and Minolta chroma meter. *Archives of Dermatological Research* 1993;285:278–282.

8. Fullerton A, Fischer T, Lahti A, Wilhelm K, Takiwaki H, Serup J. Guidelines for measurement of skin colour and erythema. A report from the Standardization Group of the European Society of Contact Dermatitis. *Contact Dermatitis* 1996;35:1–10.

9. Zografos GC, Martis K, Morris DL. Laser Doppler flowmetry in evaluation of cutaneous wound blood flow using various suturing techniques. *Annals of Surgery* 1992;215:266.

10. Diffey BL, Oliver RJ, Farr PM. A portable instrument for quantifying erythema induced by ultraviolet radiation. *British Journal of Dermatology* 1984;111:663–672.

11. Serup J, Agner T. Colorimetric quantification of erythema—a comparison of two colorimeters (Lange Micro Color and Minolta Chroma Meter CR-200) with a clinical scoring scheme and laser-Doppler flowmetry. *Clinical and Experimental Dermatology* 1990;15:267–272.

12. Takiwaki H, Serup J. Measurement of color parameters of psoriatic plaques by narrow-band reflectance spectrophotometry and tristimulus colorimetry. *Skin Pharmacology and Physiology* 2009;7:145–150.

13. Shriver MD, Parra EJ. Comparison of narrow-band reflectance spectroscopy and tristimulus colorimetry for measurements of skin and hair color in persons of different biological ancestry. *American Journal of Physical Anthropology* 2000;112:17–27.

14. Draaijers LJ, Tempelman FRH, Botman YAM, Kreis RW, Middelkoop E, Van Zuijlen PPM. Colour evaluation in scars: tristimulus colorimeter, narrow-band simple reflectance meter or subjective evaluation? *Burns* 2004;30:103–107.

15. Jain AK. Data clustering: 50 years beyond K-means. *Pattern Recognition Letters* 2010;31:651–666.

16. Jain AK, Murty MN, Flynn PJ. Data clustering: a review. *ACM Computing Surveys* 1999;31:264–323.

17. Li CH, Yuen PC. Regularized color clustering in medical image database. *IEEE Transactions on Medical Imaging* 2000; 19:1150–1155.

18. Masulli F, Schenone A. A fuzzy clustering based segmentation system as support to diagnosis in medical imaging. *Artificial Intelligence in Medicine* 1999;16:129–147.

19. Bezdek JC, Ehrlich R, Full W. FCM: the fuzzy c-means clustering algorithm. *Computers & Geosciences* 1984;10:191–203.

20. Zadeh LA. Fuzzy sets. *Information and Control* 1965;8:338–353.

21. Xu R, Wunsch DC. Clustering algorithms in biomedical research: a review. *IEEE Reviews in Biomedical Engineering* 2010;3:120–154.

22. Zanaty EA. Determining the number of clusters for kernelized fuzzy C-means algorithms for automatic medical image segmentation. *Egyptian Informatics Journal* 2012;13:39–58.

23. de Oliveira, JV, Pedrycz W. *Advances in fuzzy clustering and its applications.* New York: John Wiley & Sons, 2007.

24. Hani AFM, Nugroho H, Prakasa E, Asirvadam V, Affandi AM, Hussein SH. Soft clustering of lesion erythema for psoriasis assessment. *Journal of Investigative Dermatology* 2012;132:S66–S69.
25. Hani AFM, Prakasa E, Nugroho H, Asirvadam VS. Implementation of fuzzy c-means clustering for psoriasis assessment on lesion erythema. 2012 IEEE Symposium on Industrial Electronics and Applications (ISIEA), Kuala Lumpur, Malaysia.

3

Body Surface Area Measurement for Lesion Area Assessment

Ahmad Fadzil Mohamad Hani and Esa Prakasa

CONTENTS

3.1 Introduction

Psoriasis is one of the most common skin disorders. The disease appears as red scaling plaques that can spread on any body part [1]. In psoriasis, the immune system sends incorrect signals that speed up the growth of new skin cells. Psoriasis affects about 2%–3% of the world's population [2, 3]. The Dermatological Society of Malaysia estimates the prevalence of psoriasis in Malaysia to be 3% [4]. Since the disease cannot be totally cured, a psoriasis patient needs continuous and long-term treatment. The Psoriasis Area and Severity Index (PASI) scoring method is the gold standard for severity assessment and in determining treatment efficacy [5]. Four parameters—area (ratio of lesion area to total body surface area), erythema, scaliness, and thickness—are examined to determine the PASI score. The parameter scores from each body region (head, trunk, upper limbs, and lower limbs) are weighted and totaled to provide a PASI score ranging from 0 to 72.

For psoriasis, treatment efficacy can be considered clinically meaningful if there is a reduction of at least 75% in the total PASI score [5]. Although PASI scoring has been accepted as the gold standard, it is not used in daily practice, as scoring of all parameters is tedious and time consuming. Inter- and intrarater variation of PASI scoring by dermatologists can also occur due to the subjectivity of the PASI assessment. For PASI area assessment in particular, dermatologists usually observe lesions and their areas in four body regions. The dermatologist determines the extent of the lesion area for each region. The scores, ranging from 0 to 6, are given based on the percentage of the lesion area to the total body surface area (BSA).

3.2 PASI Area Assessment

The PASI area score is determined based on the ratio of the lesion area to the BSA of the assessed body region. PASI area assessment is a tedious task and the result may not represent the actual condition of the lesions. Currently dermatologists determine the PASI area score by roughly estimating the percentage of lesion area from visual inspection as shown in Figure 3.1. To improve the estimation, the patient is covered with a transparent plastic sheet and the borders of the lesions are drawn on the sheet. The lesion area is then determined by superimposing the plastic sheet with a grid paper. The percentage of the lesion area is then calculated by dividing the measured

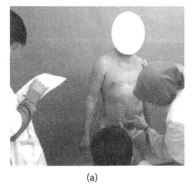
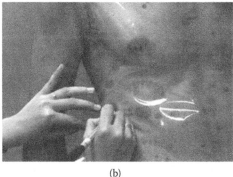

(a)   (b)

FIGURE 3.1
PASI area assessment: (a) visual assessment and (b) lesion drawing on a transparent surface.

lesion area by the BSA. However, the BSA is not measured directly but is an approximation based on the patient's height and body weight. In many cases psoriasis occurs as small and scattered lesions, which are hard for dermatologists to draw on the plastic. Moreover, it is very tedious to calculate the area of these lesions.

For an objective but simple PASI area assessment, there are two problems that need to be solved. The problems can be defined from the equation used in the PASI area parameter. The first parameter required in the PASI area equation is the BSA. Therefore an accurate measurement of BSA is important and regarded as the first problem of the PASI area algorithm. To determine lesion area accurately, the proposed algorithm should identify and segment the lesion from the normal skin area. This requirement is the second problem of the proposed algorithm. It is possible to realize an objective assessment based on accurate computation of lesion area and BSA.

Several methods are commonly used to determine BSA, including the Du Bois and Du Bois and Mosteller equations [6, 7]. These equations estimate the BSA based on body weight and height. A Lund and Browder chart can also be applied to calculate the BSA of body regions [8, 9]. The chart defines fixed percentages for each body region. Measurement accuracy can be improved by applying three-dimensional (3D) imaging methods [10]. However, specialized equipment is required to perform the 3D image acquisition [11].

Dekker et al. [12] use an eight-head body line scanner, developed by Hamamatsu Photonics (Hamamatsu City, Japan), to completely scan the entire BSA. The equipment uses near-infrared position-sensitive detectors in the sensor heads. Each sensor head contains an array of thirty-two light-emitting diodes (LEDs). The scanning mechanism moves 5 mm vertically during the cycle time to pulse the 8×32 LEDs. A scan of up to about 2000 mm, equivalent to the average human's height, gives a total of 102,400 sampled points. The full scan can be completed in approximately 10 seconds. However, the system cannot scan occluded body parts.

It is possible to reconstruct the human body surface by using silhouettes of several two-dimensional (2D) images, as shown in Figure 3.2 [13]. The images usually are obtained from a static person and image segmentation is performed to create human body silhouettes. This method is widely used to estimate a coarse 3D shape before proceeding to more sophisticated 3D shape reconstruction. The silhouette images can be easily taken if the process is performed in a controlled environment. The method might not work properly in outdoor environments, where shadow occlusions can occur. This method may be able to measure BSA, but not the lesions on the body surface, as the silhouette image does not contain the relevant information.

In previous work [14, 15], 2D imaging has been used to determine the BSA for PASI area assessment. The BSA is computed from images of only front and back views. However, it has been found that these views do not provide complete information for BSA determination. The side portions of the body regions required for full BSA measurement are not observable from these

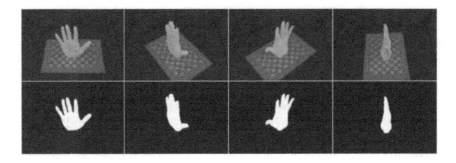

FIGURE 3.2
The sequence images that are used to reconstruct a human hand model (top) from silhouette images (bottom) [13].

views and can only be obtained from right and left views. For the inspection by dermatologists, the patient's body must be observed from all sides during the PASI area assessment.

Multiview 2D imaging has therefore been applied to acquire the BSA. The system with multiview imaging achieved reasonably good performance for the area measurement. The BSA is determined by segmenting the region belonging to the human body from the background. The scale variations in the upper limb regions and overvaluations for the surface area of the head region have been investigated to reduce unaccounted surface area, miscalculation due to overlapping areas, and scale area variations. A cross-section of the body regions is modeled by elliptical shapes rather than circular shapes to improve area measurement accuracy.

The second problem is lesion segmentation from normal skin areas. Most previous work on lesion segmentation has been based on color segmentation. The problem is difficult to resolve since people have different skin tones resulting in different hues for the psoriasis erythema. Therefore an approach that uses color dissimilarity between normal skin and the lesion is proposed to segment the lesion for various skin tones. Color parameters such as hue and chroma of normal skin and lesion samples are used to determine normal skin and psoriasis lesions.

Several researchers have conducted work on lesion segmentation. Roning et al. [16] developed a method to segment psoriasis lesion from normal skin. Grayscale images are used to distinguish the lesion from normal skin. It was assumed that the grayscale ranges of normal skin and psoriasis are different and thus an image of skin with a lesion would give a bimodal histogram. The method is initiated by dividing the image into small subimages. Threshold values are computed based on the bimodality of small subimages. An iterative method using different sizes of small subimages is performed to take into account local and global information. To test the system reliability and accuracy, the segmentation result was compared with the results of manual segmentation. The error in all cases is less than 4%. However, small psoriasis

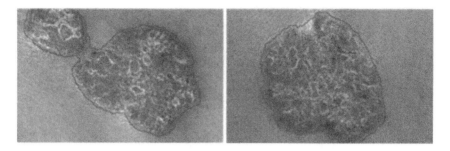

FIGURE 3.3
Segmentation results using color and texture as classifier parameters [17].

lesions are not detected by the system due to the postprocessing step. A unique postprocessing parameter is required to improve the detection ability on small psoriasis lesions.

Psoriasis lesions appear as red plaques that can be covered by silvery white scale in severe cases. Taur et al. [17] combined color and texture information to segment the psoriasis lesion. Color information is obtained from the hue and saturation components of a red, green, and blue (RGB) image. Texture information is represented by the fuzzy texture spectrum. This information is extracted from the grayscale image. The texture is based on the relative gray levels between pixels in a small subimage. The color and texture information are processed into an orthogonal subspace classifier to assign pixels into one of the two classes: psoriasis lesion or normal skin. Color and texture information from small regions of psoriasis lesion and normal skin are selected manually to obtain a training dataset. This information is used to train the classifier. Figure 3.3 shows examples of lesion boundaries produced using the method.

Taur [18] has improved the system by eliminating the manual selection of psoriasis lesion and normal skin. In the initial step, pixels of homogeneous regions are assumed to belong to either a psoriasis lesion or normal skin. A small window is passed through the image to detect homogeneity. The hue and saturation components of the skin color image and fuzzy texture spectrum of the grayscale skin image are extracted at every pass to determine region homogeneity. Several homogeneous regions may exist, however, only two types of homogeneous regions are required, psoriasis and normal skin. Thus homogeneous regions are merged according to their similarity until there are only two types of homogeneous regions. Once homogeneous regions are detected, color and texture are extracted from those regions to train a neuro-fuzzy classifier. Examples of homogeneous region detection and segmented lesion are shown in Figure 3.4.

Gomez et al. [19] analyzed RGB images using linear stepwise discriminant analysis to find suitable features for lesion–skin segmentation. Trichromatic values and their logarithms are used in the analysis. From the study, the best linear combination was given by the green and blue bands. The contrast between

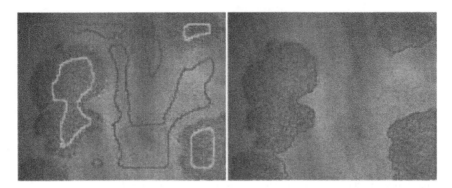

FIGURE 3.4
Typical detection result of homogeneous regions (left). The result of lesion segmentation (right) [18].

the psoriasis lesion and normal skin increases in the blue–green image. Maletti uses the blue–green image along with the red-band image to register and align a digital image of psoriasis lesions within and between treatment sessions [20]. The blue–green image is used to segment the lesion from normal skin, while the red-band image is used during the combined registration and alignment stage. However, segmentation using the blue–green image is problematic when dealing with shadowed areas. Shadows appearing at the edges of human limbs are mostly misclassified as lesion, as shown in Figure 3.5.

Jailani et al. [21] apply preprocessing, filtering, thresholding and postprocessing steps in segmenting psoriasis lesions. The preprocessing step consists of intensity adjustment and histogram equalization. A Gaussian filter, disk filter, and median filter are used for the filtering steps. The thresholding step is not clearly defined in their paper. The postprocessing step consists of

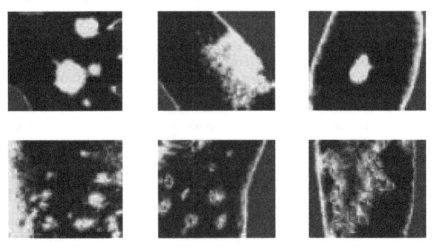

FIGURE 3.5
Segmented lesion from a blue–green image [20].

closing, opening, and filling holes. This step removes noise as well as small lesions. Thus the result concentrates only on big lesions. There are five types of psoriasis: plaque, guttate, inverse, pustular, and erythrodermic. A normalization technique is applied on RGB and grayscale images to distinguish between plaque, guttate, and erythrodermic lesions [21]. The color and gray components of psoriasis lesions are normalized by the color and gray components of normal skin from the same patient. Blue and gray components provide significant information to discriminate the three types of psoriasis with a significant mean difference at 0.05 levels.

Since skin lesions can appear in a wide variety of colors, segmentation based on a selected color is not effective. Fortunately a lesion can be detected by color gradations from its surrounding normal skin. This motivated Xu et al. [22] to use color differences based on the CIELAB color space in segmenting skin cancer. In CIELAB, a color is represented by three parameters: L^*, a^*, and b^*. L^* represents the degree of lightness, a^* represents the degree of greenness to redness, and b^* represents the degree of blueness to yellowness. Euclidean distance in the CIELAB color space is linear with perceptual color differences. To obtain color samples, the lesion is positioned in the center of the image. A sample of normal skin color is extracted automatically from the four corners of the image. Median of L^*, a^*, and b^* values are calculated from the sample to characterize normal skin color. The median is chosen rather than the mean in order to reduce the bias of skin color due to the appearance of hair on normal skin. The color difference from normal skin color is then determined. A high color difference is found in the lesion region and a small color difference is found in normal skin, as shown in Figure 3.6.

Twenty images consisting of six atypical lesion images, seven benign lesion images, and seven melanoma lesion images were analyzed. Two surgeons, one dermatologist, and one bioengineer were involved in this research in

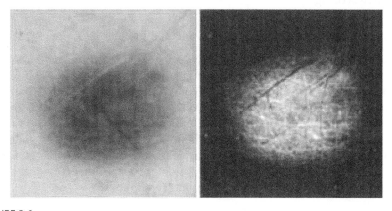

FIGURE 3.6
An example of a skin lesion color image (left). A grayscale image obtained from the color difference between normal skin and the lesion (right) [22].

order to segment the lesion manually from the image. Segmented lesions obtained manually were compared with those obtained using the automated method. The sums of the two areas that did not overlap, *a*, and the sums of the two areas, *b*, are calculated. The error is represented as a ratio of the two areas, $r = a/b$. Thus the error ranges between 0 and 1. Variability between experts is also calculated using the above formula. It was found that the overall variation between experts is slightly higher than the variation of the automated method with the four experts.

In summary, there is a need to develop an objective psoriasis lesion area assessment that avoids any subjectivity of the dermatologists' visual inspection. Since the PASI area score is determined based on the ratio of the lesion area to the BSA, there is a need to determine both the BSA and lesion areas accurately and objectively. In determining the BSA accurately, full scans or multiple views of the human body are necessary. The body area (i.e., the region of interest [ROI]) needs to be segmented effectively from the background, and in the case of multiple views, any overlapping area needs to be accounted for. From the segmented body areas, the lesion areas are then extracted from the normal skin for the PASI area assessment.

3.3 Development of the PASI Area Assessment Method

In PASI area assessment, 2D surface images of the subject's body are acquired from several views. This multiview approach is used to ensure the surfaces of all body areas are obtained. Using color data, the BSA and lesion areas are segmented from the surface images to obtain the PASI score. This section presents the methods and validation of BSA and lesion determination and PASI area scoring.

3.3.1 Image Acquisition of 2D Body Surface Images

To acquire complete 2D body surface images, surface images from four different views—anterior, posterior, right, and left—are taken. The recommended poses of the subject to ensure complete coverage of all body surfaces are shown in Figure 3.7. The views from the four poses enable the generation of a 360° representation of the human body surface.

During image acquisition (and dermatologist assessment), the subject is required to wear only disposable underwear, and for female subjects, a brassiere. Once the total body surface and lesion areas are determined, the ratio of psoriasis lesions to BSA can be calculated as follows:

$$\text{Lesion area}/\text{BSA } (\%) = \frac{A_{L,Front} + A_{L,Back} + A_{L,Right} + A_{L,Left}}{BSA_{Front} + BSA_{Back} + BSA_{Right} + BSA_{Left}} \times 100\% \quad (3.1)$$

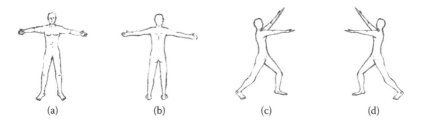

FIGURE 3.7
Body poses: (a) anterior (front) view, (b) posterior (back) view, (c) right view, and (d) left view.

Segmentation of BSA from the background and segmentation of lesions from normal skin are performed in the CIELAB color space, where L^* represents lightness, a^* is the degree of redness–greenness, and b^* is the degree of yellowness–blueness, as illustrated in Figure 3.8.

The CIELAB color space is used because it is perceptually linear with the human visual system and actual hues can be represented in the a–b plane independent of L, which is affected by ambient light. A green background has been found to be effective for segmenting the subject's body surface from the background. This is because red is the most dominant hue of human skin [23], with only small amounts of green hues. Therefore pixels due to skin color tend to have positive values in the a^* band of the CIELAB color space, whereas pixels that belong to the background tend to have negative values. The histogram on the a^* band shows two modes with clear separation, thus allowing an intensity threshold to separate the background areas and body surface (or ROI) (human skin including psoriasis lesions). Otsu's method is applied to find the threshold value.

Figure 3.9 illustrates segmentation of the upper body region. The original RGB color image (Figure 3.9(a)) is converted into the CIELAB color space

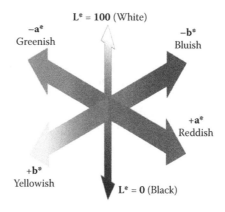

FIGURE 3.8
CIELAB color space.

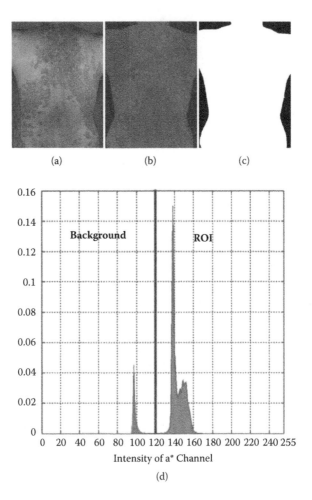

(a)   (b)   (c)

(d)

FIGURE 3.9
(a) The original RGB color image. (b) The a^* band image. (c) A binary image. (d) A histogram of the a^* band image for the ROI–background threshold.

in which the a^* band image (Figure 3.9(b)) is used. Here, the a^* values are scaled into 8-bit gray scale and thresholded to form the a^* band binary image (Figure 3.9(c)). In this example, the threshold value found using Otsu's method is 120. The histogram of the a^* band image is shown in Figure 3.9(d). It clearly shows two modes, where the background pixels are represented by the green area and the ROI is an orange color.

3.3.2 Area Segmentation of Psoriasis Lesions from 2D Body Surface Images

The process of PASI area assessment is shown in Figure 3.10. Once the background has been excluded from the image by the ROI segmentation

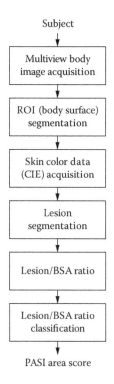

FIGURE 3.10
The process of PASI area assessment.

process, the next step is to segment psoriasis lesions from the normal skin. Psoriasis can appear in a wide variety of colors, as it is dependent on the color of the patient's normal skin and the severity of the psoriasis lesion. Therefore segmentation based on a particular color for normal skin will not be effective for all cases, as patients can have different skin tones for their normal skins.

Figure 3.10 The process of PASI area assessment. However, psoriasis lesions should be recognizable from their color dissimilarity with normal skin for any type of subject. Color dissimilarity is represented by the color difference in the CIELAB color space. The color of any pixel can be represented by its lightness (L'), hue (h_{ab}), and chroma (C_{ab}) values in the CIELAB color space. Hue (h_{ab}) and chroma (C_{ab}) are calculated using equations (3.2) and (3.3).

$$h_{ab} = \tan^{-1}\left(\frac{b^*}{a^*}\right) \tag{3.2}$$

$$C_{ab} = \sqrt{a^{*2} + b^{*2}} \tag{3.3}$$

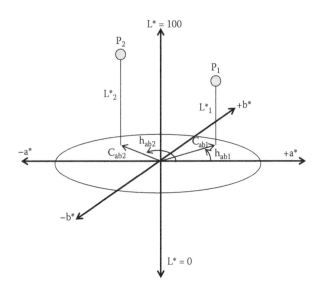

FIGURE 3.11
Color components for two colors in the CIELAB color space.

The color difference (ΔE) between two colors in the CIELAB color space is determined by

$$\Delta E = \sqrt{\Delta L^{*2} + \Delta h_{ab}^2 + \Delta C_{ab}^{*2}} \tag{3.4}$$

ΔL^* is the lightness (L^*) difference between two colors, Δh_{ab} is the hue difference between two colors, and ΔC^*_{ab} is the chroma difference between two colors. For example, Figure 3.11 shows two pixels having different colors (P_1 and P_2). P_1 is represented by L^*_1, h_{ab1}, and C_{ab1}, whereas P_2 is represented by L^*_2, h_{ab2}, and C_{ab2}. The terms ΔL^*, Δh_{ab}, and ΔC^*_{ab} are calculated as follows:

$$\Delta L^* = |L^*_1 - L^*_2|, \ \Delta h_{ab} = |h_{ab1} - h_{ab2}|, \ \Delta C^*_{ab} = |C^*_{ab1} - C^*_{ab2}| \tag{3.5}$$

Since the objective of this work is to segment a psoriasis lesion from its surrounding normal skin, only chrominance information is incorporated in the modified color difference formula, as follows:

$$\Delta E' = \sqrt{\Delta h_{ab}^2 + \Delta C_{ab}^{*2}} \tag{3.6}$$

By excluding L^*, the varied skin lightness for a person does not affect the color difference determination between the lesion and normal skin. The two colors (lesion and normal skin) are represented by two points on the hue–chroma plane, as shown in Figure 3.12. The color difference ($\Delta E'$) can thus

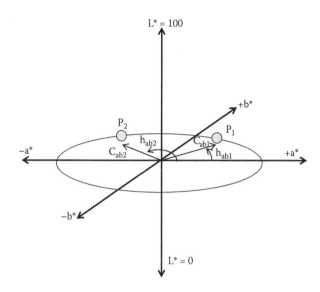

FIGURE 3.12
Color difference components between two pixels by excluding L^*.

be used as a parameter to determine whether a pixel belongs to normal skin or lesion. It is essentially the Euclidean distance between two pixels in the hue–chroma plane.

Classification based on Euclidean distance is then performed to automatically assign pixels belonging to normal skin or psoriasis lesion. Initially the user defines the sample areas of normal skin and lesion. The hue–chroma centroids of normal skin and lesion are then calculated based on this manual sampling. The centroids are calculated from selected samples of normal skin and psoriasis lesion, which are segmented manually from each image, giving an equal number of pixels. The hue–chroma centroids of normal skin and lesion are calculated by applying equations (3.7) and (3.8). Variables h_i and c_i denote the hue and chroma values of pixel i, respectively. The total number of sample pixels is represented by N. Therefore normal skin and lesion are defined by two centroids: $(h_0, c_0)_{normal}$ and $(h_0, c_0)_{lesion}$.

$$h_0 = \frac{1}{N} \sum_{i=1}^{N} h_i \tag{3.7}$$

$$c_0 = \frac{1}{N} \sum_{i=1}^{N} c_i \tag{3.8}$$

Figure 3.13 displays the separation of normal skin and lesion sample in the hue–chroma plane. The lesion data points are more scattered than the

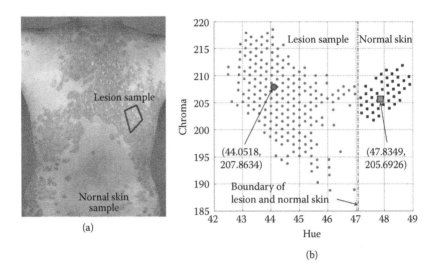

(a)

(b)

FIGURE 3.13
(a) Normal skin and lesion samples extracted from a color image of the trunk region. (b) Scatter plot of normal skin and lesion samples on the hue–chroma plane.

normal skin data points and the sample area of normal skin is larger. This occurs because the color variance of the lesion is larger compared with the normal skin. The variation is caused by redness of the lesion and the scales on the lesions, as can be seen in Figure 3.13(a).

Figure 3.14 shows the technique that can accurately segment the lesion from normal skin. Figure 3.14(a) is the ROI of the human skin and Figure 3.14(b) is the lesion area. Both images are overlaid to obtain the human skin image

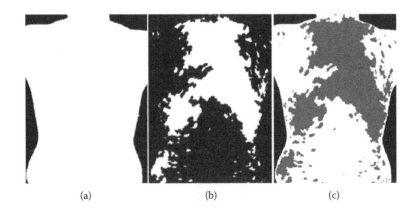

(a)   (b)   (c)

FIGURE 3.14
Lesion segmentation by applying the clustering method on the hue and chroma parameters of a 2D color image.

covered with lesion areas. The overlay result is given in the right image of Figure 3.14(c).

Once the BSA and lesion area are obtained, the area percentage of the psoriasis lesion can be calculated:

$$\text{Area percentage} = \frac{\text{Lesion area (pixels)}}{\text{BSA (pixels)}} \times 100\% \tag{3.9}$$

In the above example, the area percentage equals $(41{,}241/105{,}761) \times 100\% = 38.99\%$.

3.3.3 Correcting for Overlapped Areas in BSA Measurement

The subject's body surface images that are obtained from the front, right, left, and back viewpoints contain some amount of overlap. The overlapped areas from these multiview images have to be determined and excluded from the BSA measurement. The overlapped areas occur near the boundaries of the ROI and its background. To estimate the proportion of overlapped area, a cylindrical object with elliptical cross section is used to model the human body [24]. A cylindrical shape is considered for the cross-section model because the cross-section shape of body regions approximates an ellipse. Figure 3.15 shows cross-sectional magnetic resonance imaging (MRI) images

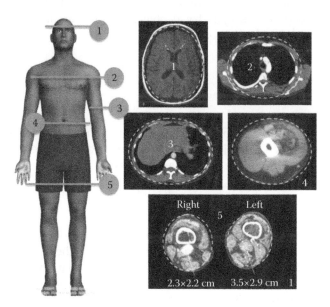

FIGURE 3.15
The cross-sectional shapes for the body regions; R1 is human body (n.d.); R2 is head (n.d.); R3 is trunk (n.d.); and R4 is arm and limb link (n.d.) [28,29,30].

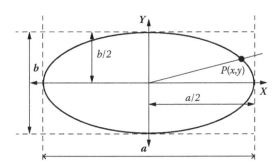

FIGURE 3.16
An ellipse with major axis *a* and minor axis *b*.

of several human body regions. The proportions of overlapping area in the horizontal and vertical directions depend on the ellipse shape. The shape is defined by major and minor axes of the ellipse. As shown in Figure 3.15, the differences in the major and minor axes at cross sections 1, 2, and 3 are large. The differences are small in cross sections 4 and 5.

The nonoverlapping areas for the multiview images can be determined by subtracting the acquired image with the estimated overlapped area. Here, the front and back views are set to be parallel with the horizontal direction. As mentioned above, the human body region is represented by a cylindrical object. The derivation steps of overlap proportion are detailed below.

The cross section of a cylindrical object is approximated by using the ellipse equation. An ideal ellipse, as shown in Figure 3.16, is formulated as

$$\frac{x^2}{\left(\frac{a}{2}\right)^2} + \frac{y^2}{\left(\frac{b}{2}\right)^2} = 1 \tag{3.10}$$

In this formulation the ellipse origin is (0, 0) and the major axis is the horizontal axis (*x*-axis). The major axis and the minor axis are denoted *a* and *b*, respectively. Equation (3.10) can be rewritten as

$$\frac{x^2}{a^2} + \frac{y^2}{b^2} = \frac{1}{4} \tag{3.11}$$

The total surface is calculated by accumulating the nonoverlapping areas from four different views (front, back, left, and right). The overlapping areas must be excluded in the BSA determination. The measured area is considered as the circumference of a rectangle that can be fitted inside the ellipse area, as shown in Figure 3.17. Therefore the ellipse circumference is approximated by the rectangle circumference. The highest accuracy is achieved when the circumference difference between the rectangle and the ellipse is a minimum.

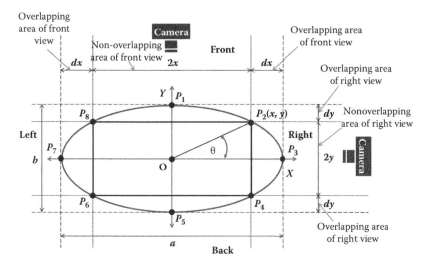

FIGURE 3.17
Overlapping and nonoverlapping areas of front and right views.

In a digital image, the ellipse curvature is seen as a flat surface. For example, in the front view acquisition, the arc $P_7P_8P_1P_2P_3$ is acquired as a flat surface P_7P_3. To obtain the nonoverlapping surface, the overlapping surfaces are determined. Table 3.1 lists the actual surface area and the measured area for each view.

From Table 3.1, the circumference of ellipse $P_1P_2P_3P_4P_5P_6P_7P_8$ is approximated by the circumference of rectangle $P_2P_4P_6P_8$. The circumference equations of the ellipse and rectangle can be written as

$$C_{\text{ellipse}} = \pi\left(\frac{3}{2}(a+b) - \sqrt{\frac{10ab}{4} + \frac{3}{4}(a^2 + b^2)}\right) \qquad (3.12)$$

$$C_{\text{rectangle}} = 4(x+y) \qquad (3.13)$$

Ramanujan's equation is used to determine the ellipse circumference [25], where θ is the boundary angle that is used to separate nonoverlapping and

TABLE 3.1

Accuracy Calculation of a Cylindrical Object

View	Actual Surface	Overlapping Surfaces	Nonoverlapping Surface	Nonoverlapping Approximation
Front	Arc $P_7P_8P_1P_2P_3$	Arc P_2P_3 (dx) and arc P_7P_8 (dx)	Arc $P_8P_1P_2$	Line P_8P_2
Back	Arc $P_3P_4P_5P_6P_7$	Arc P_3P_4 (dx) and arc P_6P_7 (dx)	Arc $P_4P_5P_6$	Line P_4P_6
Right	Arc $P_1P_2P_3P_4P_5$	Arc P_1P_2 (dy) and arc P_4P_5 (dy)	Arc $P_2P_3P_4$	Line P_2P_4
Left	Arc $P_5P_6P_7P_8P_1$	Arc P_5P_6 (dy) and arc P_8P_1 (dy)	Arc $P_6P_7P_8$	Line P_6P_8

overlapping areas and P_2 is the point in the ellipse that corresponds to angle θ (see Figure 3.17). The boundary angle is set to achieve the highest accuracy on surface measurements. This boundary angle depends on the ellipse shape that is characterized by the size of the major and minor axes. To determine the maximum circumference of the rectangle, equations (3.11) and (3.13) must be combined to cancel one of the independent variables, either x or y. Here equation (3.11) is modified to obtain y as a function of x. The modification result can be expressed as

$$y = b\sqrt{\frac{1}{4} - \frac{x^2}{a^2}} \tag{3.14}$$

Equation (3.14) is then substituted into equation (3.13):

$$C_{\text{rectangle}}(x) = 4\left(x + b\sqrt{\frac{1}{4} - \frac{x^2}{a^2}}\right) \tag{3.15}$$

The first derivative of rectangle circumference is used to determine the (x, y) coordinate of maximum circumference. The equation of the first derivative is set equal to zero, as written in equation (3.16):

$$\frac{dC_{\text{rectangle}}}{dx} = \frac{d}{dx}\left(4x + 4b\left(\frac{1}{4} - \frac{x^2}{a^2}\right)^{\frac{1}{2}}\right) = 4 + \frac{4b}{2}\left(\frac{1}{4} - \frac{x^2}{a^2}\right)^{-\frac{1}{2}}\left(\frac{-2x}{a^2}\right) = 0 \tag{3.16}$$

Equation (3.16) is then simplified to find x:

$$x = \frac{a^2}{2\sqrt{a^2 + b^2}} \tag{3.17}$$

We then substitute equation (3.17) into equation (3.14) to remove the variable x in equation (3.14):

$$y = b\sqrt{\frac{1}{4} - \frac{\left(a^2/2\sqrt{a^2+b^2}\right)^2}{a^2}} = b\sqrt{\frac{1}{4} - \frac{a^2}{4(a^2+b^2)}} = \frac{b}{2}\sqrt{\frac{b^2}{(a^2+b^2)}} = \frac{b^2}{2\sqrt{a^2+b^2}} \tag{3.18}$$

To determine x, the value of the major and minor axes (a and b) are required. By referring to Figure 3.17, the major axis is the length of P_7OP_3, whereas the minor axis is the length of P_1OP_5. The major axis can be calculated from either the front or back view images. Left or right view images might be selected to determine the minor axis. The boundary angle (θ) is then determined from the tangent relation between y and x. Equations (3.14) and (3.18) are used to substitute variable y and x, respectively. In summary, the general equation of the boundary angle can be expressed as

$$\theta = \arctan\left(\frac{y}{x}\right) = \arctan\left(\frac{b^2/2\sqrt{a^2 + b^2}}{a^2/2\sqrt{a^2 + b^2}}\right) = \arctan\left(\frac{b^2}{a^2}\right) \tag{3.19}$$

In the case where the major axis is twice the minor axis ($a = 2b$), the boundary angle is

$$\theta = \arctan\left(\frac{b^2}{(2b)^2}\right) = \arctan = \left(\frac{1}{4}\right) = 14.0362° \tag{3.20}$$

Figure 3.17 shows that dx can be obtained by subtracting x from $a/2$. For the vertical axis, dy also can be determined by subtracting y from $b/2$. By applying dx and dy values, the proportion of overlapping area can be formulated as follows:

$$\frac{dx}{a} = \frac{\left(\frac{a}{2} - x\right)}{a} = \frac{1}{2} - \frac{x}{a} \tag{3.21}$$

$$\frac{dy}{b} = \frac{\left(\frac{b}{2} - y\right)}{b} = \frac{1}{2} - \frac{y}{b} \tag{3.22}$$

Equations (3.17) and (3.14) are substituted into equations (3.21) and (3.22), respectively, to determine the proportions of overlapping area at the horizontal and vertical axes.

$$\frac{dx}{a} = \frac{1}{2} - \frac{x}{a} = \frac{1}{2} - \frac{a}{2\sqrt{a^2 + b^2}} \tag{3.23}$$

$$\frac{dy}{b} = \frac{1}{2} - \frac{y}{b} = \frac{1}{2} - \frac{b}{2\sqrt{a^2 + b^2}} \tag{3.24}$$

By applying the same case, $a = 2b$, the proportions of overlapping area are found to be

$$\frac{dx}{a} = \frac{1}{2} - \frac{2b}{2\sqrt{(2b)^2 + b^2}} = \frac{1}{2} - \frac{2b}{2\sqrt{4b^2 + b^2}} = \frac{1}{2} - \frac{1}{\sqrt{5}} = 5.28\% \tag{3.25}$$

$$\frac{dy}{b} = \frac{1}{2} - \frac{b}{2\sqrt{(2b)^2 + b^2}} = \frac{1}{2} - \frac{b}{2\sqrt{4b^2 + b^2}} = \frac{1}{2} - \frac{1}{2\sqrt{5}} = 27.64\% \tag{3.26}$$

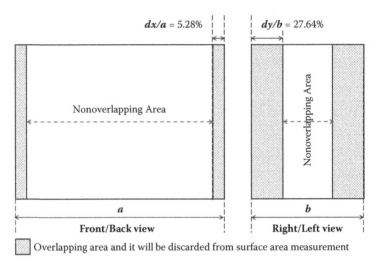

FIGURE 3.18
Nonoverlapping area of the front/back and right/left views.

From these results, it can be shown that the size of the overlapping area depends on the ellipse dimensions—the major and minor axes. The boundary angle θ is specified to reduce scale variations on the measured area. The variations are less if the surface is more flat and parallel to the camera lens. The overlapping portion of the front and back views is proportionally related to the specified boundary angle. A larger boundary angle will increase the overlapping areas of the front and back views. Since the overlapping area will be discarded, this setting can discount a large area with lesser scale variations in the BSA measurement. If the boundary angle decreases, the overlapping areas in the right and the left views will be increased. This means the counted nonoverlapping areas of right and left views will be smaller. This condition will produce no significant effect if the areas of the right and left views are small with high scale variations. The total body surface is computed by combining nonoverlapping areas from four different views. For each view, the surface area is totaled by adding pixel by pixel of nonoverlapping surface. The percentage of overlapping area is applied to correct the measurement result. Figure 3.18 depicts the nonoverlapping area of the surface image acquired from front/back and right/left views. The overlapping areas obtained in both views are not considered in the total body surface measurement.

3.3.4 Validation of Overlapping Areas in BSA Measurement

Validation of overlapping areas is performed by using a cylindrical object having an elliptical cross section as a body model. The model was wrapped with a brown tape material of known area (9509 mm²) and was used as the reference area in the validation. The actual area of each view cannot be

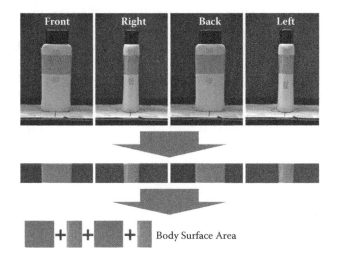

FIGURE 3.19
Four views of a cylindrical object for validating overlapping areas.

calculated directly without considering the overlapping boundaries in each view. Figure 3.19 shows the model surface was imaged from front, right, left and back views. The tape image was then extracted from the original image. Background and unwanted overlapping areas are then excluded from the BSA measurement.

The BSA of plaster material is calculated using all image views. The overlapping areas are discounted in the calculation, as listed in Table 3.2. The boundary angle θ is set to 14.0362°. As mentioned in a previous subsection, this boundary angle can give the highest accuracy in BSA measurement. At this angle, the overlapping proportions at the horizontal (dx/a) and vertical (dy/b) axes are 5.28% and 27.64%, respectively. The overlapping area reduction

TABLE 3.2

Accuracy Calculation of a Cylindrical Object

View	BSA (mm²)	Percentage of Overlapped Area for One Side (%)	Percentage of Overlapped Area for Two Sides (%)	BSA Minus Overlapping Areas (mm²)	Actual Area (mm²)	Error (mm²)
Front	4009.01	5.28%	10.56%	3585.66		
Back	3868.11	5.28%	10.56%	3459.64		
Right	2212.59	27.64%	55.28%	989.47		
Left	2135.85	27.64%	55.28%	955.15		
Total	12,225.55			8989.92	9509.00	519.08

is applied on two model boundaries of each view. Therefore the overlapping proportions are multiplied by two.

By using a known area as a reference, the accuracy of the BSA algorithm is determined, where A_{BSA} is the total BSA minus overlapping area and A_r is the actual area determined by performing manual measurement. From Table 3.2, the accuracy of the BSA algorithm is calculated as follows:

$$\text{Accuracy} = \left(1 - \frac{|A_r - A_{BSA}|}{A_r}\right) \times 100\% = \left(1 - \frac{|9509.00 - 8989.92|}{9509.00}\right) \times 100\% = 94.54\%$$

(3.27)

The error is due to the scale variation curve surface of the cylindrical object. The accuracy proves that the photography process and BSA algorithm are able to image the actual BSA of a cylindrical object. The boundary angle θ depends on the elliptical shape of the object cross section. The percentage of overlapping areas for the front and back views is proportional to the angle. If the angle increases, the overlapping areas of the front and back views will increase. This means a larger area with lesser scale variations will be discounted in the BSA measurement. If the angle decreases, the overlapping areas of the right and left views will increase. This condition can reduce the proportion of area with more scale variation (area at side surfaces). This reduction will also decrease the accuracy since the surface area of the side views cannot be neglected. As plotted in Figure 3.20, the accuracy against the boundary angle is fitted with a fifth-order polynomial ($R^2 = 1$) as follows:

$$f(x) = 3.80 \times 10^{-10} x^5 - 1.22 \times 10^{-7} x^4 + 1.53 \times 10^{-5} x^3 - 0.0009x^2 + 0.02x + 0.82$$

(3.28)

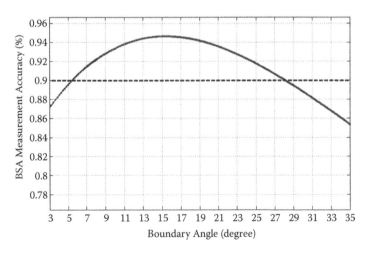

FIGURE 3.20
Plot of BSA measurement accuracy versus boundary angle.

The first derivative of equation (3.28) is used to determine the boundary angle that can give the highest accuracy. Finally, the maximum accuracy that can be achieved is at a boundary angle of 15.4007°. By comparing this boundary angle with the boundary angle that is predicted from the previous calculation, the difference is 15.4007° − 14.0362° = 1.3645°. The accuracies decrease at angles that are smaller or greater than 15.4007°.

3.3.5 Validation on a Medical Mannequin

Validation of the PASI algorithm for BSA measurement was performed by using a medical mannequin. The manual wrapping method was used to obtain the reference BSA for validating the PASI algorithm. The head, trunk, upper limb, and lower limb regions of the mannequin are wrapped as shown in Figure 3.21(a). The BSA of a body region was determined by calculating the paper areas used to cover the mannequin.

To determine the BSA using the imaging method, four images acquired from four different views were segmented into twenty-three images of various body regions, as listed in Table 3.3. The mannequin poses in the BSA images are displayed in Figure 3.21(b). The BSAs of the body regions were determined separately by applying overlapping area correction, as mentioned in a previous subsection. The head region acquired from the back view is not considered in the PASI area assessment because normally the skin surface is covered by hair.

The measurements obtained from these two methods—manual and imaging—are summarized and compared in Table 3.4.

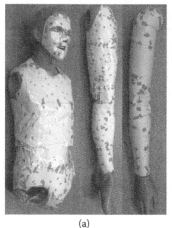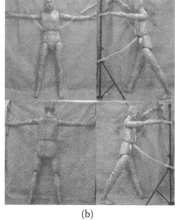

(a)　　　　　　　　　　　(b)

FIGURE 3.21
(a) Manual measurement using wrapping paper on the mannequin surface. (b) Multiple views of the mannequin and the poses.

TABLE 3.3

Selection of Images of Body Regions From Four Views of the Mannequin

Body region	Number of Images			
	Front View	**Back View**	**Left View**	**Right View**
Head	1	—	1	1
Left upper limb	1	1	1	1
Right upper limb	1	1	1	1
Trunk	1	1	1	1
Left lower limb	1	1	1	1
Right lower limb	1	1	1	1
Total	6	5	6	6

The BSA measurements for the lower limb and trunk regions achieve good accuracies (>90%). These results are achieved because (1) the ellipse cross-section model is a good approximation of the cross sections of the trunk and lower limb, and (2) for both body regions, a large proportion of the body surfaces are flat. Flat surfaces minimize any scale variation in the imaged area.

The BSA values tend to be smaller than the actual values because a fixed percentage of overlapped area is applied to all body parts. As shown in Figure 3.15, the cross-section shapes of the body regions are not uniform. The accuracy of the head region has been improved from 66.63% to 83.82% [26] by segmenting the head into three subregions from three different views. Mouth and eyelid landmarks (labial commissure and lateral commissure) are used as references to segment the face area. The accuracy for the upper limbs is relatively low due to scale area variations. The reference scales are not located at the same distance from the camera lens. To improve the accuracy of the upper limb region, the upper limb located farthest from the camera is rescaled to the upper limb nearest the camera. From the reference tapes of both upper limbs, it is known that the ratio between these two is 1.19. By using this ratio for the rescaling, the accuracy of measurements for the upper limb region is improved from 85.22% to 87.72% [26].

TABLE 3.4

Accuracy Calculation of a Cylindrical Object

Region	Manual Method (mm²)	Imaging method (mm²)	Accuracy (%)
Lower limbs	485,112.93	474,463.22	97.80%
Trunk	498,960.00	461,065.72	92.41%
Upper limbs	252,080.07	221,113.09	87.72%
Head	25,181.14	30,042.00	83.82%

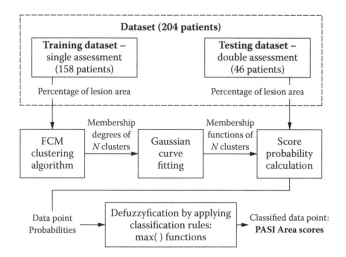

FIGURE 3.22
Development of FCM clustering for scoring the PASI area.

3.3.6 Fuzzy c-Means Clustering for PASI Area Classification

Fuzzy c-means (FCM) clustering is categorized as a soft classifier that enables a data point to become a member of more than a single cluster [27]. With this feature, FCM can resolve the clustering problems that are usually found at cluster boundaries. By applying FCM, the classification system can be built based on the actual training dataset. The statistical parameters of the training dataset (mean and standard deviation) are extracted from the training process. The development of the clustering system for PASI area assessment is described by the process flow shown in Figure 3.22. The available dataset was divided into two sets, the training and testing datasets. The training dataset was obtained from 158 patients. These images were obtained from a single assessment. Conversely, the images collected from double assessments of forty-six patients were used as the testing dataset. This dataset separation was conducted to ensure there was no mixed dataset between the training and testing datasets. The training dataset was applied to the FCM algorithm to determine the membership degrees of four clusters of PASI area scores. The clustering was applied separately for each body region. Clustered membership degrees were then fitted by applying Gaussian curve fitting to the clustered dataset. Four membership functions were created to represent score probability for any area values. To test the clustering system, a measured area was given to the four membership functions. These functions were then used to compute probability degrees of the measured area to the score clusters. From this computation, four probabilities were determined for each input measured area. In the defuzzyfication stage, the PASI area score was determined by finding the maximum probability of the score clusters. The PASI area score was selected if its probability was considered higher than the other score clusters.

3.4 Results and Analysis

3.4.1 ROI Segmentation in the PASI Area Algorithm

ROI segmentation accuracy is presented in this section. To determine accuracy, ROI segmentation obtained from the PASI area algorithm was compared with the results of manual segmentation. This manual segmentation was performed by applying the object selection tool provided by Adobe Photoshop. The results were used to determine segmentation correctness of the ROI segmentation algorithm. A total of forty-eight whole body images from twelve psoriasis patients were used to test the algorithm. These images were used to represent four types of skin tone variations (dark, brown, light brown, and fair). There were three patients for each skin tone group. Four images of four different views were collected for each patient. Therefore this setting provided a total of forty-eight images to be used in the evaluation of ROI segmentation. Figure 3.23 describes the steps of the ROI segmentation

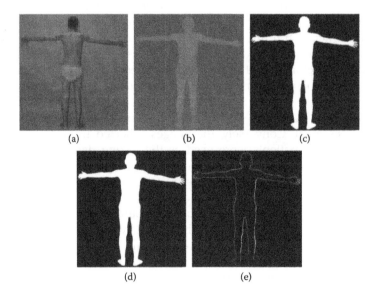

FIGURE 3.23
ROI segmentation of the back view from patient 70. (a) Original RGB color image. (b) The a* band image. (c) A binary image. (d) A binary image from manual segmentation. (e) An unmatched pixel image.

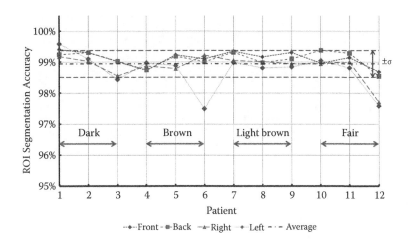

FIGURE 3.24
Plot of segmentation results from the PASI area algorithm versus manual segmentation.

process. An original RGB image (Figure 3.23(a)) was converted into the CIELAB color space. An intensity threshold was applied to the a^* band image (Figure 3.23(b)) to obtain a binary image of the patient's body (Figure 3.23(c)). A reference binary image (Figure 3.23(d)) and the ROI segmentation result were then compared by applying an exclusive or XOR operation to the image pixels. The comparison result was a binary image, as shown in Figure 3.23(e). The unmatched pixels were represented by maximum intensity, whereas a minimum intensity was used to describe the matched pixels—either skin surface or background.

From forty-eight analyzed images, the accuracies of ROI segmentation are summarized in Figure 3.24. The patients were classified into four skin tone groups: dark (patients 1, 2, and 3), brown (patients 4, 5, and 6), light brown (patients 7, 8, and 9), and fair (patients 10, 11, and 12). The average accuracy was found to be $98.94\% \pm 0.43\%$ ($\bar{x} \pm \sigma$). High accuracies of ROI segmentation can be achieved for all measurements regardless of the subject skin tone. Most of the accuracies lie within $\bar{x} \pm \sigma$. Lower accuracies were found for patients 6 and 12, images that were acquired from side (right and left) views.

Inaccuracies are caused by errors on the boundary of the ROI and the background. This can be caused by the appearance of a dark shadow on the background due to the subject's body. The shadow appears darker if the subject is close to the background. The coverage area occupied by the subject during the right/left view acquisition is always larger than front/back view. As an example, for the right view, the subject's right shoulder is located nearer to the camera, whereas his left shoulder is close to the background.

Figure 3.25 compares the ROI segmentation results of patient 6 for the front (Figure 3.25(a)) and left (Figure 3.25(b)) views. The accuracy obtained

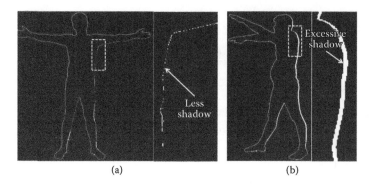

(a)   (b)

FIGURE 3.25
The images of unmatched pixels obtained from (a) front and (b) left views.

for the front view is greater than the accuracy for left view. As presented in Figure 3.25(b), excessive shadow can increase the number of unmatched pixels. The shadow can also be found in Figure 3.25(a), however, since the shadow amount is less, the number of unmatched pixels is reduced. The accuracy achieved from the front view (99.10%) is greater than the left view (97.49%). To minimize the shadow, the subject should be positioned further from the background.

3.4.2 Lesion Segmentation in the PASI Area Algorithm

The accuracy of lesion segmentation was also investigated. The accuracy was found by comparing the automatic segmentation results against manual segmentation. Manual segmentation is obtained by tracking and selecting the lesion boundaries. This selection was assisted by using the object selection tool in Adobe Photoshop, and the results were used as a reference in evaluating the lesion segmentation algorithm. A total of thirty-seven lesion samples with skin tone variations—fair, light brown, brown, and dark skin tone— were segmented in this evaluation. Comparisons between manual segmentation and the results of the PASI area algorithm are plotted in Figure 3.26.

From Figure 3.26 it can be seen that all of the PASI lesion segmentation results are near the reference line ($y = x$ line). Moreover the plot can be modeled by a linear regression model with a very high R^2 ($R^2 = 0.996$) and the fitting line is closely located with the reference line. This indicates that the PASI lesion segmentation algorithm is accurate and precise.

3.4.3 Lesion Area Classification for PASI Area Scoring

The PASI area algorithm was applied to the images of psoriasis patients, with 204 registered psoriasis patients from the Dermatology Department at Hospital Kuala Lumpur involved in the clinical study. A total of 158 patients

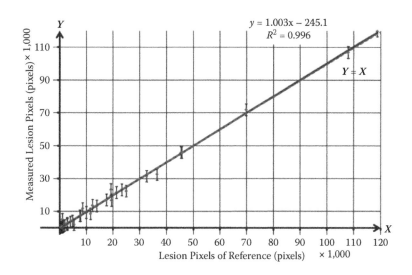

FIGURE 3.26
Plot of segmentation results from the PASI area algorithm versus manual segmentation.

provided lesion areas for developing the FCM classifier, whereas the areas from forty-six patients were used to test the classification system. Figure 3.27 shows images of four poses performed by patients that were used by the PASI area algorithm.

FCM clustering was applied to classify the lesion area into the PASI area score groups [27]. Here, the PASI area algorithm was applied to the separate body regions. Membership functions of the PASI area scores were built based on the clustering results. The clustered lesion areas with membership degrees were fitted by Gaussian curve fitting to determine membership functions of the PASI area scores.

Figure 3.28 depicts the membership functions used in the PASI area scoring of trunk region. There are six membership functions that are used to represent six grades of PASI area scores. A measured lesion area, symbolized

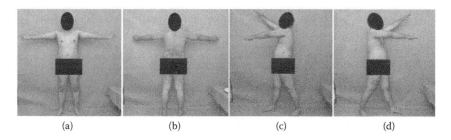

(a)   (b)   (c)   (d)

FIGURE 3.27
Four multiviews and poses of a patient: (a) front, (b) back, (c) left, and (d) right.

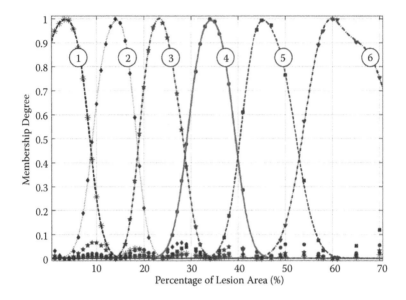

FIGURE 3.28
Membership functions of the PASI area for scoring the trunk region.

by line $x = A_L$, might cross at more than one membership function. The final score was assigned to the membership function with the higher membership degree. For example, if the measured lesion area intersects two membership functions, the lesion area was assigned to the score cluster with the higher membership degree. However, if the membership degrees of those clusters are equal, then the lesion area was classified to the higher PASI area score. If we let $x = 30\%$, according to the membership function plots, the membership degrees of this area are defined to be 0.20 and 0.64 for score 3 and 4, respectively. The lesion area is then classified as score 4 because the membership degree of score 4 (0.64) is higher than score 3 (0.20). Lesion areas collected from forty-six patients were tested by applying the FCM classifier. The PASI scores from two consecutive measurements were used to test the assessment objectivities. Two lesion area values were obtained consecutively. The testing stage requires lesion areas with measurement differences less than one standard deviation (σ) of area difference.

3.4.4 Evaluation on PASI Area Agreement

Agreement of two dermatologists and double assessments were evaluated based on the kappa coefficient. The agreement evaluations were conducted on PASI area scores for each body region. The agreement of dermatologists was evaluated by using PASI area scores for 204 patients, whereas agreement of double assessments was analyzed from the lesion areas of 46 patients. Table 3.5 lists the kappa coefficients for each body region.

TABLE 3.5

Kappa Coefficients for Body Regions

Patient	Body Region	Dermatologist 1 Versus Dermatologist 2		First Assessment Versus Second Assessment	
		N (region)	kappa	Test data	kappa
1	Head	23	0.49	11	0.76
2	Upper limbs	200	0.81	30	0.81
3	Trunk	198	0.83	28	0.85
4	Lower limbs	38	0.93	33	0.72
Total		459	0.82	102	0.80

For the upper limb and trunk regions, dermatologist and PASI area algorithm assessments resulted in similar and high kappa coefficients. Compared with dermatologists, the PASI area algorithm resulted in higher kappa agreement coefficients for the head region (0.76 versus 0.49) but lower agreement for the lower limb region (0.72 versus 0.93). In the developed classifier, six disagreements were found from thirty-three assessments of lesion area in the lower limb region. Most of disagreement cases (four of six) in the double assessments had lesion area differences less than 4%. These assessment results indicate the sensitiveness of the developed algorithm. Generally, a high kappa coefficient (≥0.80) indicates the algorithm is highly reliable and can be applied for assessing the lesion area of psoriasis.

3.5 Conclusion

A method of PASI area assessment has been presented in this chapter. The method consists of three main stages. The first stage is BSA segmentation, followed by lesion area determination and lesion area classification for determining PASI area scores. A validation study on BSA measurement was conducted to validate the accuracy of the developed method. A medical mannequin was used to model human body surfaces. Multiview imaging was applied to acquire the whole BSA of the patient. The method was validated with high accuracies on BSA determinations. The accuracies for the various body regions are 97.80% (lower limbs), 92.41% (trunk), 87.72% (upper limbs), and 83.82% (head). In the final stage, the FCM clustering algorithm was applied to the measured lesion area. From the FCM implementation, membership functions of lesion areas for PASI area scoring were determined. The final score was decided by comparing membership functions that were crossed with a certain measured lesion area. To evaluate the scoring reliability, the area score agreements from double assessments were analyzed. The kappa coefficients of the PASI area algorithm were not less than 0.72 for all body regions (head, 0.76; upper limbs, 0.81; trunk, 0.85; lower limbs, 0.72). The algorithm has higher kappa coefficients compared with dermatologist assessments for the head and trunk regions. Both assessment methods have equal kappa coefficients for the upper limb region. A lower kappa coefficient for the lower limb region was found due to system sensitivity. Overall the kappa coefficient of the developed method is 0.80, and this can be categorized as substantial agreement. This result shows that the developed PASI area algorithm is highly reliable and can be applied in PASI area assessment.

Appendix: Matlab Code

```
%% Load an image
close all;
clc, clear all

path = 'C:\Users\...\area_assessment\';

fileName = '6_trunkS.jpg';

I = imread([path,fileName]);
rgbImage = I;

% Create transformation function from RGB to Lab space
rgbToLab = makecform('srgb2lab');
LabMat = applycform(rgbImage,rgbToLab);

%% Extract L,a, b components
L = LabMat(:,:,1);
a = LabMat(:,:,2);
b = LabMat(:,:,3);

% Apply Otsu method to obtain threshold
level = graythresh(a);
% Use thershold value to segment body surface area from background
humanSkin = im2bw(a,level);
humanSkin = imfill(humanSkin,'holes');

% Convert a and b into double data type
dbl_a = double(a);
dbl_b = double(b);

% Calculate h (hue) from a and b
[mH,nH] = size(a);
for i = 1:mH
for j = 1:nH
aV = dbl_a(i,j);
bV = dbl_b(i,j);
hDeg = atand(abs(bV/aV));

        %% quadrant 1
if and(aV> = 0,bV> = 0)
h(i,j) = hDeg;
end;
```

```
          %% quadrant 2
if and(aV<0,bV> = 0)
h(i,j) = 180 - hDeg;
end;
          %% quadrant 3
if and(aV<0,bV<0)
h(i,j) = 180 + hDeg;
end;
          %% quadrant 4
if and(aV> = 0,bV<0)
h(i,j) = 360 - hDeg;
end;

end;
end;

% Calculate C (chroma) from a and b
C = sqrt(dbl_a.^2 + dbl_b.^2);

% Collect sample of normal skin
figure('Name','Select normal sample');
imshow(rgbImage);
[normArea,xN,yN] = roipoly;
close('Select normal sample');

% Collect sample of lesion
figure('Name','Select lesion sample');
imshow(rgbImage);
[lesArea,xL,yL] = roipoly;
close('Select lesion sample');

% Determine boundary of sampling area (normal skin)
min_xN = round(min(xN));
max_xN = round(max(xN));
min_yN = round(min(yN));
max_yN = round(max(yN));

% Extract h and C informations at sampling area (normal skin)
selcNormAreaBW = normArea(min_yN:max_yN,min_xN:max_xN);
selcNormAreaH = h(min_yN:max_yN,min_xN:max_xN);
selcNormAreaC = C(min_yN:max_yN,min_xN:max_xN);

% Calculate centroids of h and C at sampling area (normal skin)
[iN,jN] = find(selcNormAreaBW = =1);
N = length(iN);
sumNormDataH = 0;
sumNormDataC = 0;
for k = 1:N
sumNormDataH = sumNormDataH + selcNormAreaH(iN(k),jN(k));
sumNormDataC = sumNormDataC + selcNormAreaC(iN(k),jN(k));
end;
normCentH = sumNormDataH/N;
normCentC = sumNormDataC/N;

% Determine boundary of sampling area (lesion)
min_xL = round(min(xL));
max_xL = round(max(xL));
min_yL = round(min(yL));
max_yL = round(max(yL));

% Extract h and C informations at sampling area (lesion)
selcLesAreaBW = lesArea(min_yL:max_yL,min_xL:max_xL);
```

```
selcLesAreaH = h(min_yL:max_yL,min_xL:max_xL);
selcLesAreaC = C(min_yL:max_yL,min_xL:max_xL);

% Calculate centroids of h and C at sampling area (lesion)
[iL,jL] = find(selcLesAreaBW = =1);
N = length(iL);
sumLesDataH = 0;
sumLesDataC = 0;
for k = 1:N
sumLesDataH = sumLesDataH + selcLesAreaH(iL(k),jL(k));
sumLesDataC = sumLesDataC + selcLesAreaC(iL(k),jL(k));
end;
lesCentH = sumLesDataH/N;
lesCentC = sumLesDataC/N;

% Create a matrix for segmented image
[M,N] = size(a);
finalImg = zeros(M,N);

% Segment lesion from normal skin
[ihS,jhS] = find(humanSkin = =1);
N = length(ihS);
for k = 1:N
deltaH_toNorm_Sq = (h(ihS(k),jhS(k)) - normCentH)^2;
deltaC_toNorm_Sq = (C(ihS(k),jhS(k)) - normCentC)^2;
deltaE_toNorm = sqrt(deltaH_toNorm_Sq + deltaC_toNorm_Sq);

deltaH_toLes_Sq = (h(ihS(k),jhS(k)) - lesCentH)^2;
deltaC_toLes_Sq = (C(ihS(k),jhS(k)) - lesCentC)^2;
deltaE_toLes = sqrt(deltaH_toLes_Sq + deltaC_toLes_Sq);

    % If the colour difference close to lesion is smaller then
    % consider the pixel as lesion
ifdeltaE_toLes<deltaE_toNorm
finalImg(ihS(k),jhS(k)) = 1;
end;

    % If the colour difference close to normal is smaller then
    % consider the pixel as normal skin
ifdeltaE_toLes> = deltaE_toNorm
finalImg(ihS(k),jhS(k)) = 0;
end;

end;

% Smooth the area of segmented result
finalImg = imfill(finalImg,'holes');
SE = strel('rectangle',[3 3]);
finalImg = imerode(finalImg,SE);
lesionOnHumanSkin = not(xor(humanSkin,not(finalImg)));

% Show the images
% subplot(131),imshow(rgbImage); title('Original image');
% subplot(132),imshow(humanSkin); title('Body surface area');
% subplot(133),imshow(lesionOnHumanSkin); title('Segmented lesion');

[M,N] = size(rgbImage);
N = N/3;
redLesion(:,:,1) = zeros(M,N);
redLesion(:,:,2) = zeros(M,N);
redLesion(:,:,3) = zeros(M,N);
for i = 1:M
for j = 1:N
```

```
if and(humanSkin(i,j) = =1,finalImg(i,j) = =1)
redLesion(i,j,1) = 1;
redLesion(i,j,2) = 0;
redLesion(i,j,3) = 0;
end;
if and(humanSkin(i,j) = =1,finalImg(i,j) = =0)
redLesion(i,j,1) = 1;
redLesion(i,j,2) = 1;
redLesion(i,j,3) = 1;
end;

end;
end;
% subplot(133),imshow(redLesion); title('Segmented lesion');

subplot(121),imshow(rgbImage); title('Original image');
subplot(122),imshow(redLesion); title('Segmented lesion');
```

References

1. Menter A, Korman NJ, Elmets CA, Feldman SR, Lebwohl M, Lim HW, Van Voorhees AS, Beutner KR, Reva Bhushan PD. Guidelines of care for the management of psoriasis and psoriatic arthritis. *Journal of the American Academy of Dermatology* 2009;60:451–485.
2. National Psoriasis Foundation. Statistics. http://www.psoriasis.org/learn_statistics
3. International Federation of Psoriasis Associations. Facts about psoriasis. http://www.worldpsoriasisday.com/web/page.aspx?refid=130
4. Persatuan Dermatologi Malaysia. Overview of psoriasis in Malaysia. http://www.dermatology.org.my/overview_psoriasis.html
5. Feldman SR, Krueger GG. Psoriasis assessment tools in clinical trials. *Annals of the Rheumatic Diseases* 2005;64:65–68.
6. Du Bois D, Du Bois EF. A height-weight formula to estimate the surface area of man. *Proceedings of the Society for Experimental Biology and Medicine*. New York: Society for Experimental Biology and Medicine 1916:77–78.
7. Mosteller RD. Simplified calculation of body-surface area. *New England Journal of Medicine* 1987;317:1098.
8. Lund CC, Browder NC. The estimation of areas of burns. *Surgery, Gynecology & Obstetrics* 1944;79:352–358.
9. Miminas DA. A critical evaluation of the Lund and Browder chart. *Wounds UK* 2007;3:58–68.
10. Yu C-Y, Lin C-H, Yang Y-H. Human body surface area database and estimation formula. http://bodybrowser.googlelabs.com/body.html#m=l&note=&ui=l&opa=s:l,m:l,sk:l,c:l,o:l,ci:l,l:l,n:l&nav=-4.7,87.24,250&sel=p:;h:;s:;c:0;o:0
11. Treleaven P, Wells J. 3D body scanning and healthcare applications. *Computer* 2007;40:28–34.
12. Dekker L, Douros I, Buston BF, Treleaven P. Building symbolic information for 3D human body modeling from range data. *Proceedings of the Second International Conference on 3-D Digital Imaging and Modeling*. Washington, DC: IEEE, 1999:388–397.

13. Azevedo TCS, Tavares JMRS, Vaz MAP. Three-dimensional reconstruction and characterization of human external shapes from two-dimensional images using volumetric methods. *Computer Methods in Biomechanics and Biomedical Engineering* 2010; 13(3): 359–369.
14. Fadzil MHA, Ihtatho D, Affandi AM, Hussein SH. Area assessment of psoriasis lesions for PASI scoring. *Journal of Medical Engineering Technology* 2007;2007:3446–3449.
15. Ihtatho D, Fadzil MHA, Affandi AM, Suraiya HH. Area assessment of psoriasis lesion for PASI scoring. *IEEE Engineering in Medicine and Biology Magazine* 2007;1:3446.
16. Röning J, Jacques R, Kontinen J. Area assessment of psoriatic lesions based on variable thresholding and subimage classification. *Vision Interface, Canada* 1999;99:303–311.
17. Taur JS, Lee G-H, Tao CW, Chen C-C, Yang C-W. Segmentation of psoriasis vulgaris images using multiresolution-based orthogonal subspace techniques. *IEEE Transactions on Systems, Man, and Cybernetics, Part B: Cybernetics* 2006; 36(2): 390–402.
18. Taur JS. Neuro-fuzzy approach to the segmentation of psoriasis images. *Journal of VLSI Signal Processing Systems for Signal, Image and Video Technology* 2003; 35(1): 19–27.
19. Gomez DD, Carstensen JM, Ersbøll B, Skov L, Bang B. *Building an image-based system to automatically score psoriasis*. Berlin: Springer, 2003:557–564.
20. Maletti G, Ersbøll B, Conradsen K. A combined alignment and registration scheme of lesions with psoriasis. *Information Sciences* 2005; 175(3): 141–159.
21. Jailani R, Hashim H, Nasir Taib M. Normalization techniques for psoriasis skin lesion analysis. *Asian Conference on Sensors and the International Conference on New Techniques in Pharmaceutical and Biomedical Research*. Washington, DC: IEEE, 2005:151–153.
22. Xu L, Jackowski M, Goshtasby A, Roseman D, Bines S, Yu C, Dhawan A, Huntley A. Segmentation of skin cancer images. *Image and Vision Computing* 1999; 17(1): 65–74.
23. Maletti G, Ersbøll B, Conradsen K. A combined alignment and registration scheme of lesions with psoriasis. *Information Sciences* 2005;175:124.
24 Wang J, Hihara E. Human body surface area: a theoretical approach. *European Journal of Applied Physiology* 2004;91:425–428.
25. Cohen D, Lee TB, Sklar D. *Precalculus*. Belmont, CA: Brooks Cole Publishing, 2010.
26. Fadzil MHA, Prakasa E, Asirvadam VS, Nugroho H, Chong CH, Affandi AM, Hussein SH. Development of body surface area measurement using multi-view imaging for psoriasis area assessment. *Journal of Australasian Physical & Engineering Science in Medicine* 2011;34:591.
27. Xu R, Wunsch DC. Clustering algorithms in biomedical research: a review. *IEEE Reviews in Biomedical Engineering* 2010;3:120–154.
28. Arm and Lower Limb. http://www.tumorlibrary.com/case/detail.jsp? image_id=2133
29. Head. http://www.neurology.org/cgi/content-nw/full/64/5/919/F137
30. Trunk. http://www.info-radiologie.ch/chest_ct.php

4

Skin Lesion Thickness Assessment

Ahmad Fadzil Mohamad Hani and Hurriyatul Fitriyah

CONTENTS

4.1 Skin

The skin is the largest organ of the human body and it protects a human being's inner organs from the environment [1]. The skin consists of three layers: from outer to inner they are the epidermis, dermis, and subcutaneous fat layer. Figure 4.1 shows the structure of the skin, in which the epidermal layer is colored red, the dermis is colored pink, and the subcutaneous fat layer is colored yellow [2].

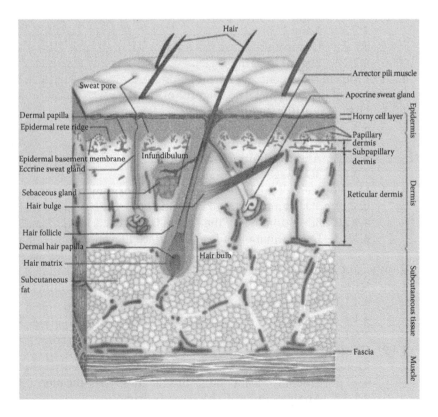

FIGURE 4.1
The structure of the skin [2].

The thickness of the epidermis is on average about 0.2 mm and consists of four layers: the basal cell layer, the suprabasal cell layer, the granular layer, and the horny cell layer. The dermis that lies beneath the epidermis is fifteen to forty times thicker than the epidermis. It consists of the papillary layer, subpapillary layer, and reticular layer. The subcutaneous layer consists of fat cells that preserve neutral fat, retain moisture, generate heat, and cushion against external physical pressure. An eruption on the skin known as a lesion can occur anywhere on the body surface and can affect some or all layers of the skin. A lesion can exhibit abnormal coloring, swelling, rash, scaling, weeping, pus, or scabbing.

4.1.1 Skin Lesion

Skin lesions are classified into three types based on the cause [3]. The first type are skin lesions due to infection by biological organisms (i.e., parasites and bacteria). The second type is the result of exposure to external environmental or physical conditions such as light, heat, cold, electricity, and pressure. Chemical substances in the body such as drugs, allergens, and toxins can cause the third type of lesion.

Skin lesions can also be classified into primary and secondary lesions based on their appearance [1, 2, 4]. Primary lesions occur in skin without any preexisting eruption and include

1. A change in color (e.g., macule, patch)
2. An elevation above normal skin (e.g., papule, plaque, nodule)
3. A serum or pus container (e.g., blister, pustule, cyst)
4. A temporary elevation (e.g., wheal)

Secondary skin lesions are eruptions from a preexisting eruption. This skin lesion type can also be due to treatments, progression of disease, or manipulation such as scrubbing and scratching. Secondary skin lesions include crusts, scales, erosions, ulcers, scars, and keloids. From the various types of skin lesions mentioned, only macules, patches, erosions, and ulcers do not show thickening of the epidermis layers. Macules and patches exhibit color changes of the skin surface, while erosions and ulcers appear as holey skin damage.

Examples of skin diseases that are related to the occurrence of skin lesions with abnormal skin thickening are listed in Table 4.1. The table also shows the change in skin structure caused by the skin lesion along with its description.

4.1.2 Psoriasis

Psoriasis is among the skin diseases that cause elevation or thickening on the epidermis layer. Psoriasis is a chronic inflammatory skin condition that

TABLE 4.1

Skin Lesions and Associated Skin Diseases

No	Skin Lesion	Skin Structure	Description	Skin Disease
1	Papule		Pimple-like elevation of skin with diameter <10 mm	Carcinoma, Lichen planus; Neurofibromatosis
2	Plaque		Pimple-like elevation of skin with diameter >10 mm	Psoriasis; Mycosis, Bowen disease
3	Nodule		Solid and round lesion with diameter of 10–20 mm	Metastatic cancer; Dermatofibroma
4	Blister		Circumscribed and elevated cavity; contains fluid (serum, blood)	Varicella, Bullous Pemphigoid; Eczema; Pemphigus
5	Pustule		Circumscribed superficial cavity of skin; contains purulent	Herpes zoster; impetigo; psoriasis; smallpox; Rosacea
6	Cyst		Dome-shaped tumorous lesion; contains liquid or solid materials	Cystic acne
7	Wheal		Rounded, flat, pale red papule or plaque	Urticaria
8	Crust		Solidified exudates (fluid, purulent) which dries on skin surface	Ecthyma
9	Scale		Abnormal white flake of stratum corneum	Psoriasis; Ichthyosis; Keratosis
10	Scar/keloid		Tissue replacement after healing of injury wound or ulcer	Keloid
11	Athropy		Thinned or a smooth wrinkled skin surface; result of aging	Discoid lupus erythematous; Lichen planus

is caused by an accelerated replacement of human skin cells [5, 6]. In normal conditions, skin cells shed and replace themselves in twenty-eight days, whereas in psoriasis it takes only four days [7]. Galen, in 133–200 AD, was the first person to use the term psoriasis, from the Greek word *psora*, which means itch. It is one of the most common skin diseases and affects about 125 million people around the world, or 2%–3% of the world's population [8]. In Malaysia, the prevalence of psoriasis is 3% [9]. A study of the outpatients in the Dermatology Department, Hospital Tengku Ampuan Rahimah, Malaysia, conducted between January 2003 and December 2005 showed that 9.5% of all outpatients were found to be suffering with psoriasis [10].

Psoriasis can occur at any stage of life, but most commonly it occurs during the second and third decades of life. It occurs in both genders, with an equal ratio of male to female sufferers and afflicts people of any race. Psoriasis is noncontagious. There is also no significant evidence that psoriasis runs in families [7]. Only a few cases have shown that a child whose parent has psoriasis may also contract psoriasis. The genetic predisposition for both identical twins to have psoriasis is 65%–75%, and 15%–20% for nonidentical twins. Psoriasis tends to appear as a combination of genetic and environmental conditions, such as stress and infection.

In clinical diagnosis, psoriasis lesions consist of erythematous, sharply demarcated and elevated plaques with silvery white scales [11]. The lesion has an irregular oval shape, varying from one to several centimeters in diameter. There are several types of psoriasis based on symptoms and appearance: plaque, pustular, guttate, inverse, erythodermic, and arthritic [8]. In general, patients only have one type of psoriasis occurring on their body at a time. About 80% of psoriasis cases are plaque psoriasis, also known as psoriasis vulgaris [6, 12]. Plaque psoriasis appears with silvery white scales on top of red patches. The silvery white scales denote skin shedding as a result of accelerated skin growth. Old scales can be shed by scratching or rubbing. The red patches are due to the increase in blood vessels to support the increased skin cell production. A larger patch is formed from several localized and smaller patches. An example of psoriasis vulgaris affecting a patient's lower back is shown in Figure 4.2.

In an aggravated psoriasis condition, there is an elongation and thickening of the rete ridges, elongation and edema of the dermal papillae, an absent or poorly formed granular layer, parakeratosis (incomplete keratinization with retention of nuclei), microabscesses in the epidermis, enlargement of the capillaries, and a mainly lymphocytic infiltrate around the subpapillary vessels [8]. Figure 4.3 shows a histology image of skin with psoriasis [2]. The image shows the thickening of the epidermis layer caused by the psoriasis.

Recent research into the causes of psoriasis has found that the change in skin growth is affected by a change in the immune system [6], as T

FIGURE 4.2
Plaque psoriasis.

lymphocytes or T cells are triggered and then become overactive. The T cell acts as if it is fighting a virus, an infection, or healing a wound. This T cell activity accelerates the skin growth, forming the thickened plaques in the epidermis layer. Table 4.2 shows the frequency of localizations of psoriasis lesions from a study of 784 respondents in the United Kingdom [13]. Psoriasis commonly occurs on the scalp, elbows, legs, knees, arms, and trunk, but can appear on any part of the human body.

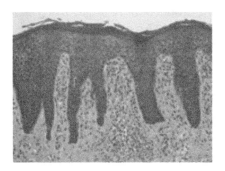

FIGURE 4.3
Histology of psoriasis [2].

TABLE 4.2

Location of Psoriasis, $n = 784$ Respondents [13]

Location	% of respondents
Scalp	79.7
Elbows	77.8
Legs	73.7
Knees	56.6
Arms	54.2
Trunk	52.6
Lower part of body	46.8
Base of back	38.0
Other	37.7
Soles of feet	12.9

4.2 Thickness Assessment of Skin Lesions

In skin lesion severity assessment, the thickness of the skin lesion is one of the most important parameters to be assessed, along with color change, distribution, eruption type, shape, elevation, and roughness. The assessment varies from tactile inspection performed by a dermatologist (subjective assessment) to an objective assessment that gives quantitative measures of lesion thickness.

4.2.1 Subjective Assessment

Tactile inspection is the basic technique used by dermatologists to assess the thickness of skin lesions in daily practice. The affected skin area is examined by touching it several times to compare the skin thickening or elevation of the lesion with the surrounding healthy skin. The dermatologist will also run his/her fingers on the lesion in different directions to confirm the assessment. Figure 4.4 shows how a dermatologist performs the tactile inspection to assess the thickness of a skin lesion using his/her fingers.

However, the dermatologist's assessment is subjective. Subjectivity can lead to inter- and intrarater variation. Interrater variation indicates different scores given by two or more dermatologists, while intrarater variation describes a different score given by one dermatologist after repeated assessments.

4.2.2 Objective Assessment

Several approaches have been developed to quantify skin lesion thickness objectively using a microscope, X-rays, ultrasound, or digital imaging technologies.

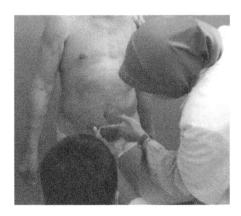

FIGURE 4.4
A dermatologist assesses the thickness of a skin lesion using tactile inspection.

4.2.2.1 Light Microscope

A light microscope is used to examine the histology image of the skin lesion, showing the subcutaneous layers [14, 15]. Figure 4.5 shows an example of a light microscope [16].

The histology image taken using a camera is transferred to a computer to obtain the thickness measurement. The epidermis thickness is determined by measuring the distance between the top and bottom of the rete. Another study used digital dermoscopy to enlarge the details of the skin structures using optical magnification [17]. Epidermal thickening due to the skin lesion occurrence is measured using computer software. Figure 4.6 shows an example of digital dermoscopy [18].

FIGURE 4.5
Olympus BX50 microscope [16].

FIGURE 4.6
Digital dermoscopy [18].

To obtain a histology image, a sample of skin tissue is taken using biopsy methods. Figure 4.7 shows how the biopsy is performed. The dermatologist marks the area of the skin and takes a sample using a punch or incision technique, as shown in Figure 4.7 [2].

A histology image can show well-defined boundaries between the epidermis, dermis, and subcutaneous fat structures. The image can also be used to assess the elongation and elevation of the skin layers in early, advanced, and later stages of psoriasis [19]. However, a biopsy is an invasive technique [14]. Dermatologists have to consider appropriate biopsy locations when dealing with cosmetically important sites.

4.2.2.2 X-ray

X-ray radiation has been widely used on patients' skin to identify different subcutaneous layers [20]. The patient's skin is imaged on a photographic

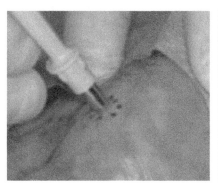
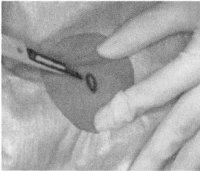

FIGURE 4.7
Skin biopsy techniques [2]: (a) punch method; (b) incision method.

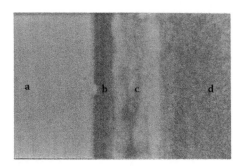

FIGURE 4.8
X-ray projections of the skin [22]: (a) Lucite block; (b) skin; (c) fat; (d) muscle.

plate and examined to determine the thickness of the lesion. The use of X-ray requires flattening a forearm against a Lucite or wood block and projecting the X-ray beam in a parallel direction [21]. A projection captured by the plate is shown in Figure 4.8 [22].

Physicians examine the X-ray image to determine epidermal skin thickness. The X-ray projection technique has a radiation hazard and this technique cannot be used too often [22]. This technique is also limited to several body parts, as the position between the projection, skin, and photographic plate has to be parallel.

4.2.2.3 Ultrasound

Ultrasound technology has been used to measure skin thickness by calculating the time needed for high-frequency sound pulses to travel from the transducer to skin layer and reflect back to the transducer [22]. The skin acoustical velocity, known as travel time, is thus related to distance. The travel time is transformed to a spatial representation of the skin's internal structure. In the measurement of skin lesion thickness, ultrasound with frequencies of 40 MHz, 20 MHz, 15 MHz, 10 MHz and 7.5 MHz is applied to the skin lesion to quantify the thickness [23–28]. Ultrasound technology is a noninvasive method and does not involve hazardous radiation. Ultrasound can be applied to any part of the body. Although it is a noninvasive technique, the epidermis, dermis, and subcutaneous fat echoes are not always easily distinguishable [22]. An A-scan plot of the ultrasound resonance from the skin sample in Figure 4.8 is shown in Figure 4.9 [22].

Another type of ultrasound resonance image is the B-scan. The image can show a cross-sectional image of the skin layers (i.e., the epidermis, dermis, and subcutaneous fat layers). Figure 4.10 shows the visual and B-scan region of a skin lesion using 35-MHz ultrasound [29].

The above three imaging modalities in skin lesion imaging—light microscope, X-ray, and ultrasound—attempt to look at epidermal thickening

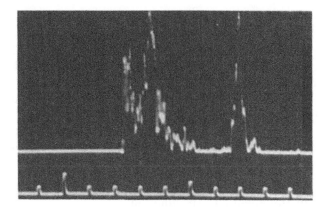

FIGURE 4.9
An A-scan plot of the ultrasound resonance of skin [22].

under the skin surface. The disadvantage of light microscope, X-ray, and ultrasound is that an expert is needed to take the images and segment the epidermis layer from other skin layers.

4.2.2.4 Digital Imaging

Instead of seeing the underlying structure of the skin, recent developments in medical imaging have led to the use of digital imaging (two-dimensional [2D] and three-dimensional [3D]) to obtain information from above the skin. 2D digital imaging has been widely used to characterize skin lesions based on color information. In psoriasis, 2D digital imaging is used to measure skin lesion area coverage [30–32] and 3D digital imaging is used to image

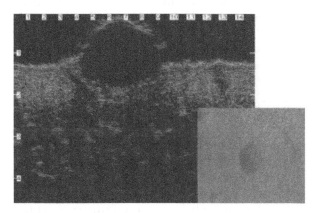

FIGURE 4.10
Visual and B-scan region of a skin lesion using 35-MHz ultrasound [29].

the surface profile of the skin lesion for skin texture characterization (e.g., roughness). It is mostly applied in cosmetic dermatology to measure the texture of cellulite and wrinkles [7, 33, 34]. Moreover, several 2D bidirectional images are captured to form a 3D image of the skin that is used for texture characterization [35].

Digital imaging is a noninvasive technique to obtain color and textural information of a patient's skin lesion. It is low cost and digital images are highly reproducible. The color and texture information of digital images can be analyzed automatically using computer systems. The digital imaging technology, especially 3D, can provide surface profiles of the lesion.

4.2.3 Problem Statement and Formulation

In daily clinical practice, dermatologists assess the thickness of skin lesions via tactile inspection. They touch and feel the alteration of the skin due to the lesion and compare it to the surrounding normal skin. Based on knowledge and experience, tactile inspection can be performed several times and in several directions on the skin lesion in order to provide a comprehensive assessment.

In the severity assessment of skin lesions, dermatologists need to use a standard assessment tool. The assessment tool gives a standard descriptor that guides the doctor in assessing the severity of the skin disease as mild, moderate, or severe based on the skin lesion's condition. Although based on a standard tactile descriptor, the tactile inspection results in a subjective assessment. It does not result in a quantitative measure of lesion thickness. Thus the outcomes can be different between dermatologists (interrater variation) and between repetitive assessments of a single dermatologist (intrarater variation).

To overcome the subjectivity and variations in thickness assessments, an objective technique is required to provide a quantitative measure of lesion thickness. Currently the objective measurement is performed by measuring the abnormal thickening of the epidermis layer caused by the skin lesion's occurrence [14, 15, 23–28]. The inner structure image of the skin is captured using digital imaging modalities such as a light microscope, scanning electron microscopy (SEM), transmission electron microscopy (TEM), high-resolution magnetic resonance imaging (HR-MRI) and high-frequency ultrasound (HFUS).

Here, an objective measurement that adapts the dermatologist's technique (i.e., tactile inspection), which only can perceive alterations of the skin lesion against the surrounding normal skin, is developed. The developed measurement algorithm has to be able to measure the thickness of the elevated lesion surface from the lesion base. The lesion base is the normal skin before the abnormal elevation occurs.

The surface of the human body is an uneven surface that poses a technical challenge. The 3D curvature in the chest area is different with the curvature in the abdomen and thigh areas. Therefore the lesion base cannot be assumed as flat but has to follow the curvature or surface trend of the surrounding normal skin in 3D.

Lesion thickness is determined by measuring the thickness of the lesion surface from the lesion base. The lesion base is the normal skin below the lesion surface. Once the lesion base is known, the thickness, which is the elevation difference between the lesion surface and the lesion base can be measured. The objective of the research was thus to develop an algorithm that can measure in vivo the thickness of the skin lesion from the determined lesion base. The second objective of the research was to determine the range of the lesion thickness for each severity level of the standard assessment tool to enable grading.

To acquire the surface profile comprising the elevations of the lesion surface and its surrounding normal skin, a 3D digital imaging modality is used. The thickness measurement is then performed on the 3D space of the acquired 3D surface profile. The abnormal elevation on the skin surface is the skin eruption or skin lesion (e.g., as a plaque psoriasis, papule, nodule, blister, pustule, cyst, wheal, crust, scale, erosion, scar, or keloid). Other forms of skin lesion, such macules and patches that show abnormal skin coloring and ulcers that show a deep erosion of skin, are excluded.

4.3 Development of a 3D Thickness Algorithm

In this section, the development of a 3D image analysis algorithm to assess objectively the thickness of skin lesions is discussed. The image acquisition setup including the 3D scanner and its settings, ambient illumination, and algorithm requirements are discussed in section 4.3.1. The design of the lesion thickness algorithm is presented in section 4.3.2. Section 4.3.3 discusses the validation process of the developed algorithm. In section 4.3.4, application of the thickness assessment algorithm in a clinical study is presented.

4.3.1 Data Acquisition Setup

For consistent quality in the 3D image data acquisition, the settings of the 3D scanner and the ambient lighting are standardized as follows.

4.3.1.1 3D Imaging Equipment

A 3D optical scanner using fringe projection is used to acquire a 3D surface image of a lesion and its surrounding normal skin. For this study, the PRIMOS 3D portable scanner shown in Figure 4.11 was used for its light weight (5 kg), portability, and fast image acquisition (<70 ms).

The specifications of the PRIMOS 3D portable scanner are listed in Table 4.3 [36].

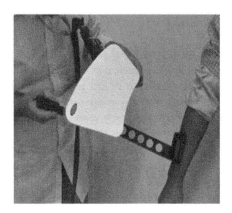

FIGURE 4.11
PRIMOS 3D portable scanner.

4.3.1.2 3D Scanner Setup

The PRIMOS 3D portable scanner consists of a light projector, a charge-coupled device (CCD) camera and a guide frame that is intended to maintain a fixed distance of 22 cm between the CCD camera and the object to be scanned (Figure 4.12). The fixed distance produces a focused 3D surface image of a standard size, 4 cm × 3 cm (320 pixels × 240 pixels, with each pixel equal to 0.0125 mm × 0.0125 mm). The guide frame is placed on the patient's skin such that the CCD camera is directly on top of the region of interest (ROI) (i.e., the camera axis is perpendicular to the skin ROI) to minimize surface occlusions.

4.3.1.3 Ambient Illumination

The PRIMOS projector lighting illumination is 1420 lux (lx) at the end of the guide frame (measured using a lux meter). The ambient lighting is set at a lower illumination relative to the projector in order to ensure a clear fringe

TABLE 4.3

Specifications of the PRIMOS 3D Portable Scanner

Specification	Unit
Measuring field	4 cm × 3 cm
Lateral resolution	63 µm
Elevation resolution	≥4 µm
Accuracy of elevation measurement*	≥6 µm
Measuring time	<70 ms

* Elevation accuracy achieved in a PRIMOS standard calibration process.

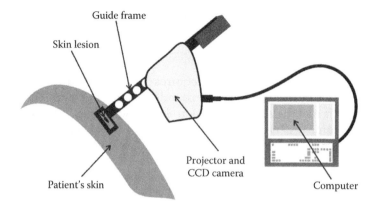

FIGURE 4.12
Schematic diagram of image acquisition using the PRIMOS 3D portable scanner.

projection pattern is formed on the skin. The ambient lighting in the acquisition room in this study was set to be around 450 lx, which can be achieved by using a fluorescent lamp.

4.3.1.4 3D Image Capture

The PRIMOS user interface is equipped with a live-view screen, as shown in Figure 4.13, and is the user interface to acquire 3D images. Referring to the

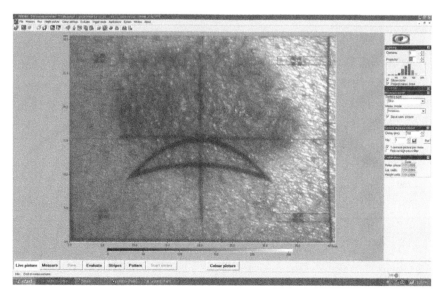

FIGURE 4.13
Live view of the PRIMOS user interface.

figure, the dark plaque is the lesion, while the lighter-tone skin is the surrounding normal skin. An umbrella-like pattern is projected onto the surface to assist in focusing the image. The acquired 3D image data is a regular grid of coordinates (x, y) with elevation (z).

The lesion thickness algorithm requires that the acquired 3D surface image consist of a single lesion surrounded by normal skin. The second requirement is that the surface elevation should not exceed the depth of field of the CCD camera to ensure the fringe pattern is always in focus throughout the image. The CCD camera has a depth of field of ±5 mm at the end of the guide frame.

An example of a properly acquired skin visual is shown in Figure 4.14(a) and the corresponding 3D skin image, $I(x, y, z)$, is shown in Figure 4.14(b).

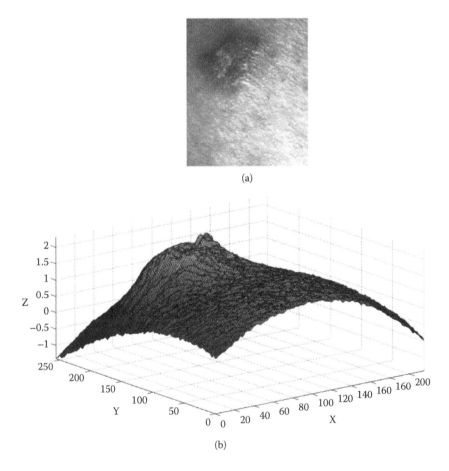

(a)

(b)

FIGURE 4.14
(a) Visual skin lesion. (b) 3D skin surface image.

4.3.2 Development of the Lesion Thickness Algorithm

From the 3D skin surface image, the elevation of the scanned skin surface can be obtained. The elevation of a lesion on the skin is measured from the base of the lesion to its top surface. However, the lesion base has to be determined first from the acquired 3D surface image. The lesion base is the normal skin before the lesion occurred and thus it is assumed to have the same surface trends as the surrounding normal skin. The lesion base is therefore estimated from the surrounding normal skin using a 3D interpolation method, hence the need to have a substantial amount of surrounding skin during image acquisition. This method has been used to estimate the surface of ulcer wounds in 3D profile to measure their volume [37]. Figure 4.15 illustrates a cross section of a skin lesion where the thickness of the lesion at any point is the elevation of the lesion surface from its base.

Since the 3D skin surface image consists of the lesion and surrounding normal skin, the surrounding normal skin is first segmented from the lesion. An automatic segmentation method using thresholding was developed to segment the normal skin from the lesion and this creates a mask, $I_m(x, y)$, from which the coordinates of the normal skin (x_n, y_n) and lesion (x_l, y_l) are known. The coordinates are then used to find the respective elevation of normal skin (z_n) and lesion (z_l) on the 3D skin surface image.

Two surfaces are thus determined, the normal skin $I_n(x, y, z)$ and the lesion $I_l(x, y, z)$. The segmented normal skin is then best fitted using a polynomial surface fitting to produce an estimated surface, $I_e(x, y, z)$. Best fitting was chosen over exact fitting because the purpose of the surface fitting is to follow the trend or curvature of the normal skin. The estimated lesion base is taken to be the best-fitted surface at the lesion coordinates (x_l, y_l). The 3D surface image $I(x, y, z)$ is then subtracted from the estimated surface $I_e(x, y, z)$ at the lesion coordinates (x_l, y_l) in order to determine surface elevations at the lesion coordinates. The surface elevations at the lesion coordinates are then averaged to obtain the lesion thickness. The process flow of the algorithm is shown in Figure 4.16 and the MATLAB code for the algorithm is given in Appendix 4.A.

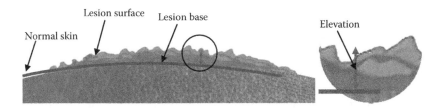

FIGURE 4.15
Thickness measurement of a skin lesion based on surface elevation.

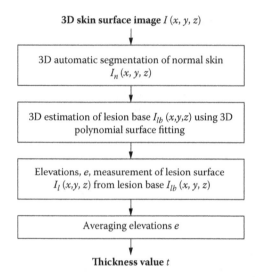

FIGURE 4.16
Algorithm to measure lesion thickness.

4.3.2.1 3D Segmentation of Normal Skin

A segmentation process to determine which pixels in the 3D skin surface image $I(x, y, z)$ belong to the normal skin and which pixels belong to the lesion was developed. Segmentation in the 3D image is performed on a corresponding 2D elevation map.

In the 3D surface image segmentation, the image is initially converted to a 2D elevation map where the elevations are represented by grayscale intensities. The range of elevation is scaled to 0–255 in the 8-bit pixel case using equation (4.1):

$$I_{2D}(x,y) = 255 \times \frac{z - \min(I(x,y,z))}{\max(I(x,y,z)) - \min(I(x,y,z))} \tag{4.1}$$

An example of image conversion from a 3D surface image to a 2D elevation map is shown in Figure 4.17.

The corresponding 2D elevation map of the 3D skin surface image in Figure 4.14 is shown in Figure 4.18.

Lesion segmentation can be performed manually by the user, but automatic lesion segmentation was developed for the lesion thickness algorithm. The segmentation in the corresponding 2D elevation map can be performed automatically using the edge detection, region growing, or thresholding methods. Here, the normal skin is segmented using the thresholding method since the intensity of the lesion and surrounding normal skin can be separated easily with certain threshold values. The edge detection method was

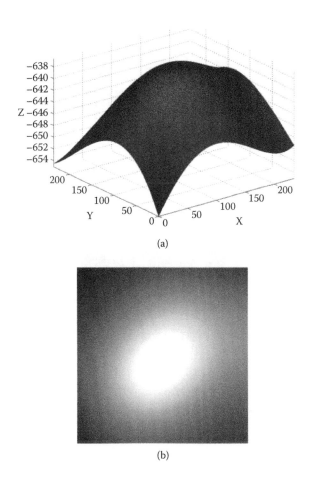

(a)

(b)

FIGURE 4.17
Conversion of a 3D surface image to a 2D elevation map: (a) 3D surface image; (b) 2D elevation map.

not chosen because the elevation gradient between the edge of the lesion and the normal skin is not large, hence the edge is not easily detected. Region growing segmentation was also not chosen because the technique requires a seed point, which has to be determined manually by the user.

The human skin surface is an undulating surface. As a result, the intensity of the elevated skin lesion on the corresponding 2D elevation map (example in Figure 4.18) cannot be differentiated from the intensity of the surrounding normal skin. The segmentation results of the 2D elevation map in Figure 4.18 using thresholding at intensities 240, 230, and 200 are shown in Figure 4.19.

As seen in Figure 4.19, skin lesions and normal skin cannot be segmented directly from the current 2D elevation map using the thresholding method. This problem can, however, be overcome. To have clear intensity differences between the lesion and normal skin pixels on the corresponding 2D elevation

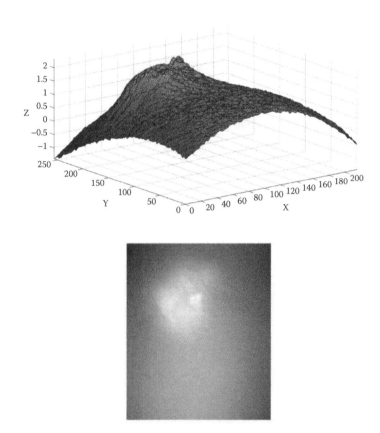

FIGURE 4.18
Corresponding 2D elevation map of a 3D skin surface image.

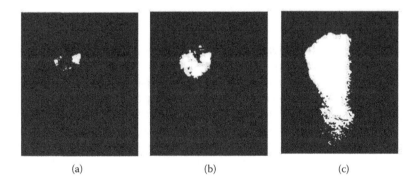

FIGURE 4.19
Segmentation of a 2D elevation map using thresholding at intensities of (a) 240, (b) 230, and (c) 200.

map, the 3D skin surface image $I(x, y, z)$ needs to be flattened (uncurved). The process used flattens the normally curved human skin surface into a flat skin surface in order to obtain only the elevated surface due to lesion thickness; the elevation of normal skin is generally near zero and the lesion will have positive elevation values.

Figure 4.20 illustrates a cross section of the 3D skin surface image from Figure 4.14(b), as indicated by the blue line. By constructing a blue line that

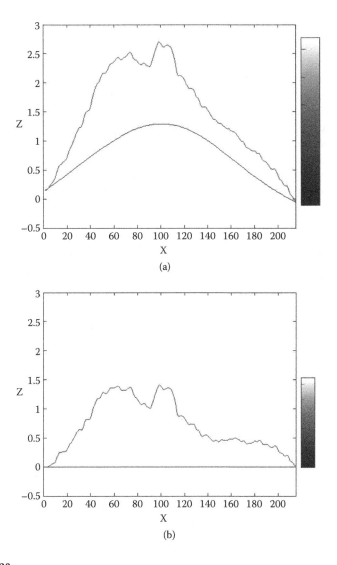

FIGURE 4.20

3D skin surface image flattening. (a) A plot showing a slice of a 3D skin surface image. (b) The elevated surface only.

resembles the curved human skin surface in Figure 4.20(a), the elevated surface due to the lesion is obtained by the flattening process (see the new blue line in Figure 4.20(b)). The grayscale intensity in Figure 4.20 shows that the flattening process decreases the range of surface elevation. Thus the elevation difference between the skin lesion and normal skin in the corresponding 2D elevation map is well defined, enabling a fixed threshold to be used.

The blue lines in Figure 4.20 are the surface trend of the curved 3D skin surface image. The surface trend, that is, the ideal surface, $I_i(x, y, z)$, is formed by filtering the 3D skin surface using a moving average filter. In 3D surface application, this is commonly performed to obtain the trend of the surface [38]. However, in the 3D skin surface image that has a concave curvature, the ideal surface $I_i(x, y, z)$ cannot estimate the trend of the surface. The applied filter tends to elevate the ideal surface to follow the elevation on the side of the surface, which is higher than the elevation at the center, hence the elevated surface cannot be obtained. This implies that automatic segmentation does not work on a concave surface. A plot showing a slice of a 3D skin surface image that has a concave curvature is shown as a red line in Figure 4.21(a), while the estimated ideal surface is shown as a blue line. The obtained elevated surface only is shown in Figure 4.21(b). Thus, for concave surfaces, the segmentation is performed manually.

The window size of the moving average filter in 3D surface applications is determined by experiment. In this developed 3D automatic segmentation, the size depends on the image size (x, y) of the 3D skin surface image. The larger the image size, the larger the window size. From experimentation, selection of the window size, n, is shown in Table 4.4.

The ideal surface $I_i(x, y, z)$ of the 3D skin surface image of Figure 4.14(b) is shown in Figure 4.22.

The 3D skin surface image is then subtracted from the ideal surface $I_i(x, y, z)$ to determine the elevated surface, $I_{es}(x, y, z)$. Figure 4.23(a) shows the vertical elevation surface $I_{es}(x, y, z)$ as the result of subtraction between the 3D skin surface image (Figure 4.14(b)) and the ideal surface (Figure 4.21), where only positive results are considered (negative values are capped to zero). Negative values sometimes exist at the edges of the surface because the filtered surface tends to be higher than the actual surface at the edges. The corresponding 2D elevation map of the vertical elevation surface is shown in Figure 4.23(b).

The spurious bright spots in the 2D elevated surface of Figure 4.23(b) are noise due to the scanning process. These should be removed as they can later be misclassified as lesion areas. To remove the spurious noise, a 15×15 median filter is applied to the vertical elevation surface. Similar to the moving average filter, there is no rule for window size in the median filter and it

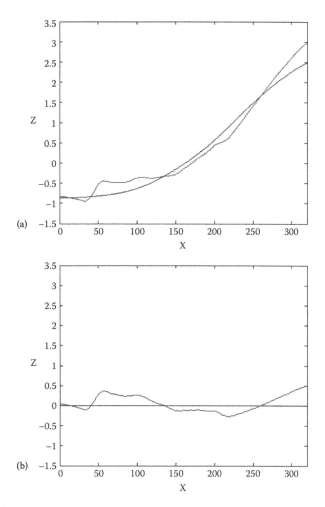

FIGURE 4.21
Concave 3D skin surface image flattening. (a) A plot showing a slice of a 3D skin surface image.
(b) The elevated surface only.

TABLE 4.4

Window Size of the Moving Average Filter

$x + y$	N
1–200	$(x + y)/8$
201–250	$(x + y)/6$
251–300	$(x + y)/4$
301–500	$(x + y)/4 + 15$
501–560	$(x + y)/3$

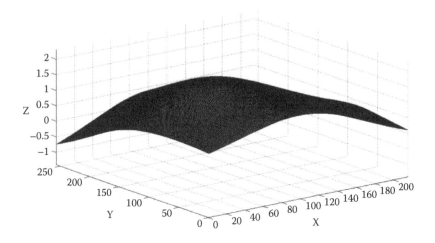

FIGURE 4.22
Ideal 3D skin surface image.

is determined by experiment. A filtered vertical elevation surface, $I_{fes}(x, y, z)$, with the corresponding 2D elevated map is shown in Figure 4.24.

In the 2D elevation map (Figure 4.24(b)), the intensity difference between the lesion and normal skin is clearly defined compared with the corresponding 2D elevation map obtained directly from the unfiltered 3D skin surface image (Figure 4.23(b)).

Normal skin areas in the 2D elevation map are segmented by thresholding. From 0 to 255 grayscale intensity, the threshold is found from observations to be 100, where intensity >100 is considered as lesion and <100 is considered normal skin. The masking process gives value 0 to the normal skin and 1 to the skin lesion using equation (4.2), resulting in a masked 2D elevation map $I_m(x, y)$.

$$I_m(x,y) \begin{cases} 0, (x,y) \in \text{normal skin} \\ 1, (x,y) \in \text{lesion} \end{cases} \tag{4.2}$$

From the masked 2D elevation map, $I_m(x, y)$, the pixels that belong to the normal skin (x_n, y_n) and lesion (x_l, y_l) are known. The masking result of the 2D elevation map in Figure 4.24 is shown in Figure 4.25.

The segmented pixels of normal skin (x_n, y_n) and lesion (x_l, y_l) are then used to find the respective normal skin surface $I_n(x, y, z)$ and lesion surface $I_l(x, y, z)$ on the 3D skin surface image as shown in Figure 4.26.

The process flow of 3D automatic segmentation of normal skin is shown in Figure 4.27.

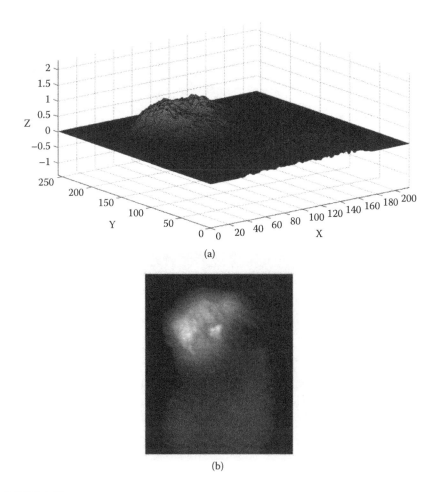

FIGURE 4.23
(a) Vertical elevation surface. (b) Corresponding 2D elevation map showing the presence of spurious noise.

4.3.2.2 3D Estimation of the Lesion Base

The segmented normal skin surface is used to estimate the lesion base, $I_{lb}(x, y, z)$. The normal skin surface in Figure 4.26(a) shows empty data points at the lesion coordinates (x_l, y_l) where the lesion base is to be estimated. The segmented normal skin surface $I_n(x, y, z)$ is then best fitted using a 3D polynomial least-squares surface fitting to construct the estimated surface or lesion base, $I_e(x, y, z)$.

The polynomial orders used in the surface fitting are third, fourth, and fifth. The second-order polynomial is not used, as it is impossible to have a flat skin

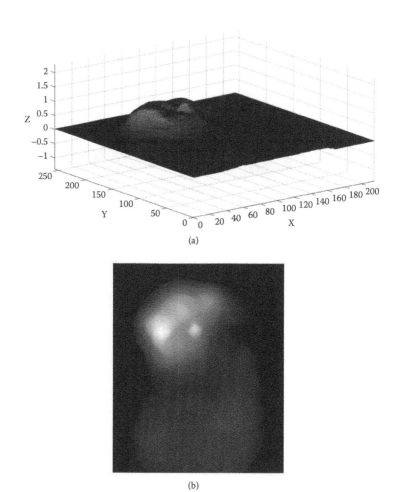

FIGURE 4.24
(a) Vertical elevation surface after median filtering. (b) Corresponding 2D elevation map.

FIGURE 4.25
Segmentation result of the 2D elevation map [51].

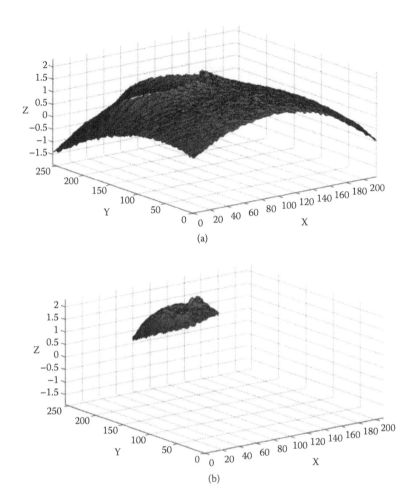

FIGURE 4.26
Segmentation result of the 3D skin surface image: (a) normal skin surface; (b) lesion surface.

surface in the scanned 3D skin surface image. The equations for the third, fourth, and fifth orders of the 3D polynomial functions are given as follows.

$$Z = f(x, y) = a_0 + a_1x + a_2y + a_3x^2 + a_4y^2 + a_5xy + a_6x^3 + a_7y^3 + a_8x^2y + a_9xy^2 \quad (4.3)$$

$$Z = f(x, y) = a_0 + a_1x + a_2y + a_3x^2 + a_4y^2 + a_5xy + a_6x^3 + a_7y^3 + a_8x^2y \\ + a_9xy^2 + a_{10}x^4 + a_{11}y^4 + a_{12}x^3y + a_{13}x^2y^2 + a_{14}xy^3 \quad (4.4)$$

$$Z = f(x, y) = a_0 + a_1x + a_2y + a_3x^2 + a_4y^2 + a_5xy + a_6x^3 + a_7y^3 + a_8x^2y \\ + a_9xy^2 + a_{10}x^4 + a_{11}y^4 + a_{12}x^3y + a_{13}x^2y^2 + a_{14}xy^3 + a_{15}x^5 \\ + a_{16}y^5 + a_{17}x^4y + a_{18}x^3y^2 + a_{19}x^2y^3 + a_{20}xy^4 \quad (4.5)$$

These functions are also listed in Table 4.B.1 in Appendix 4.B.1

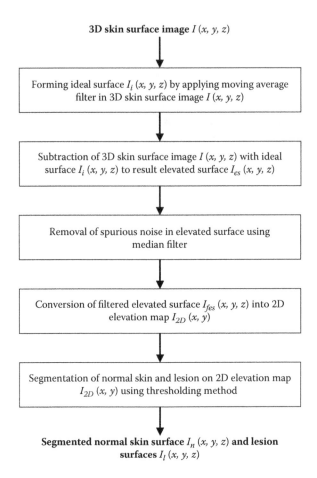

3D skin surface image $I(x, y, z)$

Forming ideal surface $I_i(x, y, z)$ by applying moving average filter in 3D skin surface image $I(x, y, z)$

Subtraction of 3D skin surface image $I(x, y, z)$ with ideal surface $I_i(x, y, z)$ to result elevated surface $I_{es}(x, y, z)$

Removal of spurious noise in elevated surface using median filter

Conversion of filtered elevated surface $I_{fes}(x, y, z)$ into 2D elevation map $I_{2D}(x, y)$

Segmentation of normal skin and lesion on 2D elevation map $I_{2D}(x, y)$ using thresholding method

Segmented normal skin surface $I_n(x, y, z)$ **and lesion surfaces** $I_l(x, y, z)$

FIGURE 4.27
Process flow of 3D automatic segmentation of normal skin [52].

As described in Appendix 4.B, there are two steps to perform the 3D polynomial least-squares surface fitting. The first step is to find the polynomial coefficient matrix [a] using equation (B.2). After obtaining the polynomial coefficient matrix [a], the estimated surface $I_e(x, y, z)$ is determined using equation (B.3). Three estimated surfaces are constructed using third, fourth, and fifth orders of polynomial surface fitting. The estimated surface chosen is the one that gives the closest fit or minimum fitting error to the segmented normal skin.

As an example, using the fourth order of polynomial surface fitting, the estimated surface $I_e(x, y, z)$ of the segmented normal skin surface $I_n(x, y, z)$ is shown in Figure 4.28(a). The lesion base is the estimated surface on lesion coordinates (x_l, y_l) as shown in Figure 4.28(b).

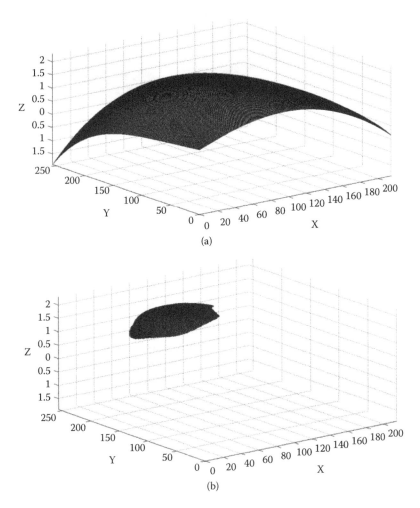

FIGURE 4.28
Surface fitting result of normal skin: (a) estimated surface; (b) lesion base.

4.3.2.3 Determination of Lesion Surface Elevation for Thickness

The lesion elevation, e, from the lesion base is obtained by subtracting the 3D skin surface image $I(x, y, z)$ and the estimated surface $I_e(x, y, z)$ at the lesion coordinates (x_l, y_l). In the 3D lesion base estimation, the elevation of the lesion base at the edges of the lesion can be slightly higher than its respective lesion surface. This is because the best fitting attempts to get the trend of a rough 3D skin surface image instead of its exact elevation values. This situation results in negative lesion elevations, which are not significant to the total lesion elevation values and are thus excluded. Figure 4.29 shows an example of negative lesion elevations at the edges of a lesion. The lesion surface

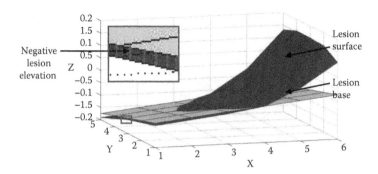

FIGURE 4.29
Negative lesion elevations at the edges of a lesion.

is colored dark gray, the lesion base is colored light gray, and the negative lesion elevations are shown as red arrows.

The thickness value (t) is determined by averaging the total (n) lesion elevations (e) using equation (4.6):

$$t = \frac{\sum_{i=1}^{n} e_i}{n} \tag{4.6}$$

Averaging the irregular elevations is chosen as an analogue to the dermatologist's thickness assessment of the skin lesion. The dermatologist presses and feels the lesion from edge to edge several times and in several directions and this is considered as averaging the irregularity of the lesion elevations. Averaging the elevations is also used to determine the thickness of irregular elevations on sea ice [39] and paper [40].

4.4 Results and Analysis

The following sections present the validation of the lesion thickness assessment method and analyses of the results from the implementation of the proposed lesion thickness algorithm in an observational clinical study entitled Objective Evaluation of Psoriasis Severity Using a New Computerized PASI Scoring System (Alfa-PASI) (NMRR ID: NMRR-09-1098-4863) to measure the thickness of plaque psoriasis lesions in vivo. The validation and analysis of the lesion thickness algorithm from segmentation to lesion base estimation to thickness assessment are discussed in section 4.4.1. Section 4.4.2 discusses the clinical study results and performance of the lesion thickness algorithm in clinical settings.

4.4.1 Validation and Analysis of the Lesion Thickness Algorithm

The 3D scanner has a certain depth of field, which is 5 mm above and below the end of the guide frame. The acquired 3D surface image, which has surface

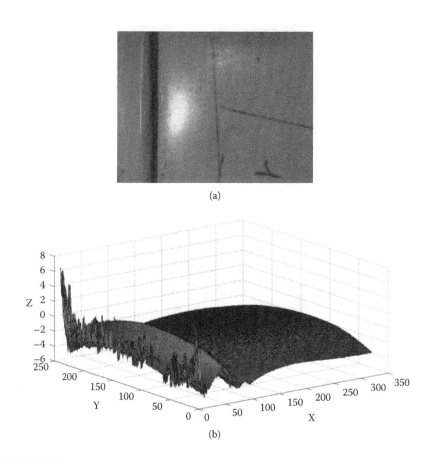

FIGURE 4.30

Out-of-focus 3D surface sample acquired from the joint of a mannequin: (a) visual; (b) 3D surface model.

elevations exceeding this range of depth of field, is considered as an out-of-focus image. Images found to be out of focus are thus excluded from the dataset. The out-of-focus images normally occur at the joints of mannequins, body folds, back of the hands, and other locations where the range of surface elevations exceeds the depth of field of the scanner. Examples of visual and 3D surface images of the out-of-focus 3D surface samples are shown in Figures 4.30–4.32, respectively.

From the total 1,110 3D surface samples acquired from life-size mannequins, only 325 in-focus 3D surface samples were used as the dataset in the validation. From the female mannequin, 168 3D surface samples were collected: head, 1; trunk, 54; upper extremities, 34; and lower extremities, 79. From the male mannequin, 157 3D surface samples were collected: trunk, 82; upper extremities, 15; and lower extremities, 60, as tabulated in Table 4.5.

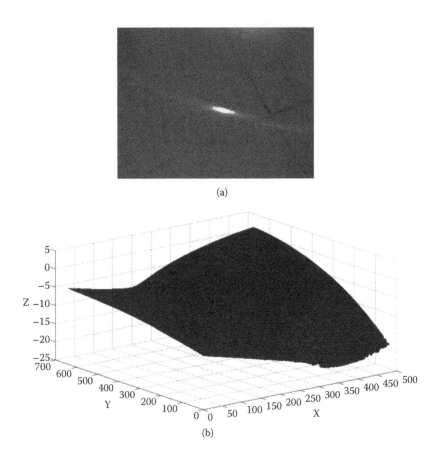

FIGURE 4.31
Out-of-focus 3D surface sample acquired from the breast fold of a mannequin: (a) visual; (b) 3D surface model.

4.4.1.1 Analysis of 3D Normal Skin Segmentation

The first validation was performed on the 3D automatic segmentation method. In this validation, the ability of the 3D automatic segmentation to segment the normal skin of 3D segmentation surfaces that consist of the lesion model added to the 3D surface samples was investigated. The segmented pixels of normal skin and lesion from manual segmentation of the lesion model were used as the reference. The result of the segmentation is 2D masking, which gives value 0 to the segmented normal skin and 1 to the lesion. The Cohen's kappa interrater agreement in equation (4.7) was used to determine the agreement strength between the 2D masking of the segmentation result and the reference.

The dataset of the 3D segmentation surfaces used in this validation refers to the selected in-focus 3D surface samples in Table 4.5. The distribution of kappa values of 325 3D segmentation surfaces is shown in Figure 4.33. The

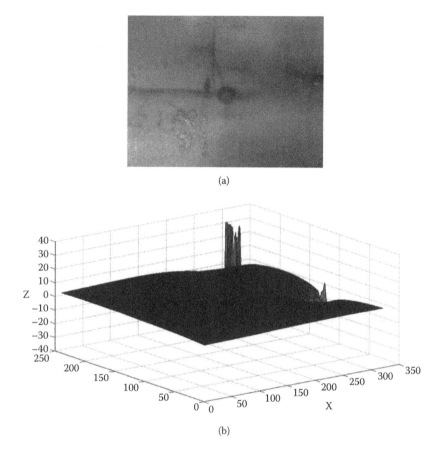

(a)

(b)

FIGURE 4.32

Out-of-focus 3D surface sample acquired from the back of the hand of a mannequin: (a) visual;
(b) 3D surface model.

TABLE 4.5

Dataset of the 3D Surface Samples

Mannequin Body Part	In-focus 3D Surface Samples	
	Female	Male
Head	1	0
Trunk	54	82
Upper extremities	34	15
Lower extremities	79	60
Total = 325	168	157

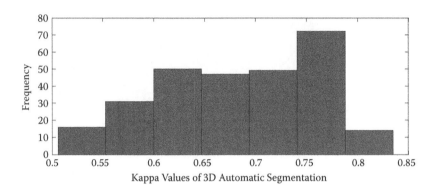

FIGURE 4.33
Kappa values of 325 3D segmentation surfaces.

automatic segmentation method was not applied to the surface samples that
have concave curvatures.

Since the kappa values are >0.60, and based on Table 4.C.1, the proposed
3D automatic segmentation can determine whether a pixel belongs to normal
skin or lesion. By performing this validation, the developed 3D segmentation of normal skin can determine the designation of each pixel to normal
skin and lesion of the 3D skin surface image with substantial segmentation
performance ($\kappa > 0.60$) for every nonconcave surface on the human body. The
segmentation on a concave surface should be performed using a manual
segmentation.

4.4.1.2 Analysis of 3D Lesion Base Estimation

The 3D lesion base estimation is an important step in the lesion thickness
algorithm, as the measured lesion thickness relies directly on how well and
correct the lesion base is constructed. The 3D surface samples $I_s(x, y, z)$ of the
validation dataset are cropped in the center using a rectangular shape. The
cropping process was tested for robustness to 25%, 50%, and 75% of the total
surface area. The 3D polynomial surface fitting was then used to estimate the
cropped center. Then the mean absolute error (MAE) between the cropped
center $I_{cc}(x, y, z)$ and the estimated center $I_{ec}(x, y, z)$ was calculated.

The distribution fitting of the MAE for 25%, 50%, and 75% cropping is
shown in Figure 4.34 as a red line, blue dashes, and green dots, respectively.

The average of 325 MAEs for 25%, 50%, and 75% cropping are listed in Table 4.6.

As expected, the error of the surface fitting method increases when the
cropping area is increased. Wider area of the cropping reduces the area of
the surrounding surface. In the 3D polynomial surface fitting, the surrounding surface is an important reference to estimate the cropped center, where
a wider area of the surrounding surface results in a more accurate estimated

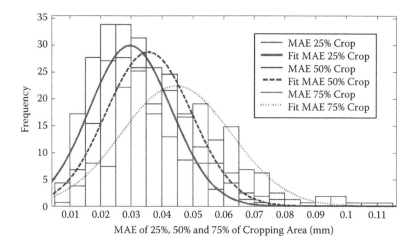

FIGURE 4.34
Mean absolute error (MAE) of 325 3D surface samples [51].

center. Thus, in the lesion thickness algorithm, a wider area of surrounding normal skin results in a more accurate lesion base.

As listed in Table 4.3 Specifications of the PRIMOS 3D portable scanner, the scanner's elevation resolution is 4 μm, thus the acceptable error tolerance is stated as 0.04 mm, which is ten times the elevation resolution. Based on Table , the acceptable error tolerance is achieved when the cropping area is no more than 50% of the total surface area. Therefore, from this validation exercise, the 3D polynomial surface fitting can estimate the lesion base within the error tolerance of 0.04 mm so long as the area of the lesion is no larger than 50% of the total area of the 3D skin surface image.

4.4.1.3 Analysis of the Thickness Algorithm

The lesion thickness algorithm was tested in the third validation process to measure the thickness of a lesion model taped on different surfaces of the

TABLE 4.6

Average MAE of Surface Fitting to Construct the Lesion Base [52]

Cropping Area (% of Surface)	Average MAE (mm)
25%	0.030 ± 0.013 ($\bar{X} \pm$ SD)
50%	0.037 ± 0.014 ($\bar{X} \pm$ SD)
75%	0.045 ± 0.019 ($\bar{X} \pm$ SD)

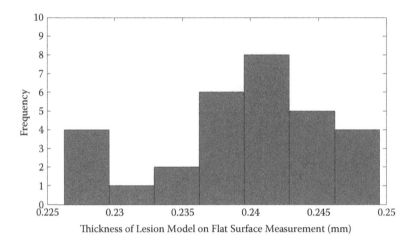

FIGURE 4.35
Thickness distributions of 30 lesion models on flat surface measurement.

human body. The lesion model uses a surgical tape size of 2.5 cm × 1.2 cm, which is 25% of the measurement area of the PRIMOS 3D portable scanner. To determine the reference thickness value of the lesion model, the thicknesses of thirty lesion models taped to a flat surface were measured using the proposed algorithm. For a flat surface measurement, the base of the lesion model is flat, hence the lesion base is constructed using the second order 3D polynomial surface fitting. The distribution of the lesion model's thicknesses on the flat surface measurement is shown in Figure 4.35.

The reference thickness value of the lesion model was then calculated as the average thicknesses of thirty lesion models on the flat surface measurement. From the thickness distribution of thirty lesion models shown in Figure 4.35, the reference thickness was found to be 0.24 mm ± 0.007 mm ($\bar{X} \pm$ SD).

After determining the reference thickness value of the lesion model, the lesion models that were taped on the life-size mannequins and scanned are analyzed using the lesion thickness algorithm to determine its thickness value. The dataset of the 3D lesion model surfaces used in this validation also refers to the selected in-focus 3D surface samples in Table 4.5. The thickness distribution of lesion models measured using the lesion thickness algorithm is shown in Figure 4.36.

From the thickness distribution of 325 lesion models shown in Figure 4.36, the error was determined by calculating the difference between the reference thickness (0.24 mm) and the average thickness of the simulation result (0.283 mm ± 0.029 mm [$\bar{X} \pm$ SD] or 0.283 mm ± 0.055 mm [$\bar{X} \pm$ 2SD]). In this validation, the error of the thickness measurement using the lesion thickness algorithm is 0.043 mm. Referring to the MAE of the 25% cropping test in the validation of the 3D lesion base estimation shown in Table 4.6, it was

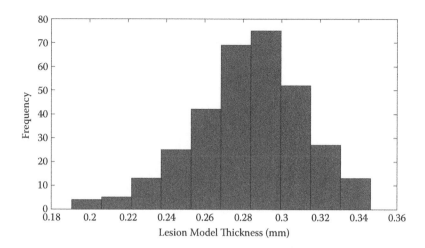

FIGURE 4.36
Thickness distributions of 325 lesion models.

confirmed that the 0.043 mm error of the lesion thickness algorithm for the 25% lesion model area is within the error range of 0.030 mm ± 0.013 mm ($\bar{X}$ ± SD) and slightly above the error tolerance of 0.04 mm. Thus the lesion thickness algorithm can determine the thickness of a lesion model with an error of 0.043 mm on the human body.

4.4.2 Clinical Study Results and Analysis of the Lesion Thickness Algorithm

4.4.2.1 Thickness Data and PASI Thickness Classification

The lesion thickness algorithm was applied in a observational clinical study to measure the thickness of plaque psoriasis lesions in vivo. A total of 713 plaque psoriasis lesions were studied from 125 male and 29 female patients with ages ranging from nineteen to seventy-four years. The patients are registered in Dermatology Department of Hospital Kuala Lumpur, Malaysia. The demographics of the patients recruited, including age, sex, and race, are shown in Figure 4.37.

In this study, two independent dermatologists used tactile inspection to assess the severity of lesion thickness and score them subjectively based on the description of the Psoriasis Area and Severity Index (PASI). Visual examples of psoriasis lesions with scores 1–4 assessed by dermatologists are shown in Figure 4.38.

After the lesion thickness was assessed by two independent dermatologists, the 3D skin surface images of the lesion along with its surrounding normal skin were captured using the PRIMOS 3D portable scanner. As designed, the 3D segmentation of normal skin in the lesion thickness algorithm is not

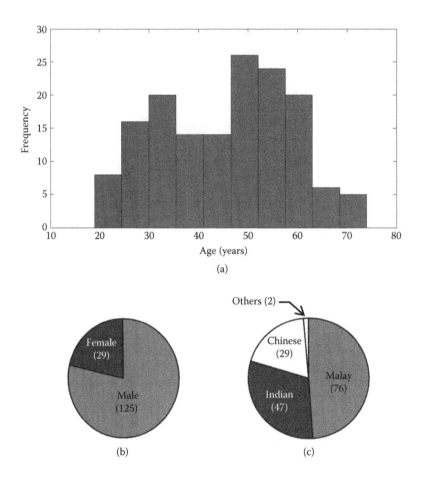

FIGURE 4.37
Patient demographics of the clinical study: (a) age; (b) sex; (c) race.

applied on concave surfaces. To deal with this, patients were asked to bend their body for chest and waist locations in order to acquire convex surface curvatures for 3D skin surface images.

The in-focus and non-concave 3D skin surface images were then analyzed using the lesion thickness algorithm to determine the thickness of the lesion. The thicknesses of 713 plaque psoriasis lesions were found to range between 0.053 mm and 0.684 mm, as shown in Figure 4.39.

The interrater agreement analysis of the dermatologists' assessments was found to have a kappa value of 0.52, indicating moderate agreement between the two dermatologists (see Appendix 4.C). However, this agreement is well below 0.60 (substantial agreement), hence the dermatologists' scores were not suitable for classification, thus unsupervised k-means clustering was used instead. The k-means clustering was applied to group the measured thicknesses of plaque psoriasis lesions into PASI thickness scores 1 to 4.

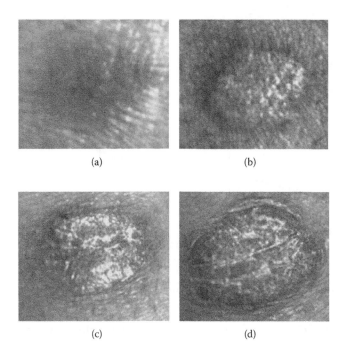

(a)   (b)

(c)   (d)

FIGURE 4.38
Plaque psoriasis lesions assessed by dermatologists with PASI scores (a) 1, (b) 2, (c) 3, and (d) 4.

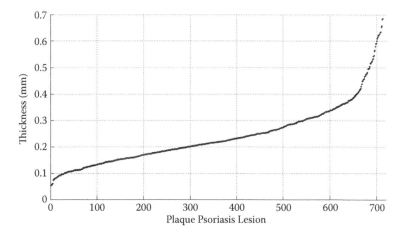

FIGURE 4.39
Lesion numbers (lesion 200 and 600) are given in the graph [51].

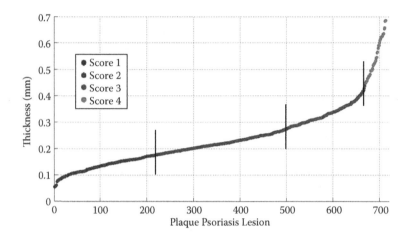

FIGURE 4.40
Thickness scores are added to the graph. The borderlines of the scores are also highlighted [51].

Given the 713 lesion thicknesses, the k-means clustering assigns each lesion thickness into four clusters based on the nearest distance to the cluster's center (centroids). The clustering results of 713 lesion thicknesses are shown in Figure 4.40. The thicknesses scored as 1 are colored dark blue, score 2 is colored green, score 3 is colored red, and score 4 is colored light blue.

The number of data points, N, centroid, and thickness range of each PASI thickness score as the result of clustering are listed in Table 4.7.

It can be seen from Table 4.7 that score 4 has the fewest data points. This condition also happens in reality, where skin lesions with severe condition are rare [41]. This is because once the symptoms appear, patients seek medical treatment and thus lesion severity scores seldom reach 4.

4.4.2.2 Analysis of PASI Thickness Scoring

The PASI thickness score centroids and ranges shown in Table 4.7 were validated to determine the performance of the classification. Since the

TABLE 4.7

Centroid and Thickness Range of the PASI Thickness Score

Score	N	Centroid (mm)	Thickness range (mm)
1	223	0.134	0.053–0.177
2	280	0.222	0.178–0.277
3	165	0.333	0.278–0.439
4	45	0.547	>0.440

TABLE 4.8

Validation Result of PASI Thickness Scoring

| Score | Centroid of Complete Dataset (mm) | Subdataset 1 | | Subdataset 2 | |
		Centroid (mm)	Difference (mm)	Centroid (mm)	Difference (mm)
1	0.134	0.134	0.000	0.134	0.000
2	0.222	0.222	0.000	0.223	0.001
3	0.333	0.329	0.004	0.337	0.004
4	0.547	0.535	0.012	0.549	0.002

classification method used is k-means clustering, the performance is determined using centroid stability analysis. The 713 lesion thicknesses were equally divided in a random fashion using the MATLAB function (crossvalind) into two subdatasets. The first dataset had 366 data points and the second dataset had 347 data points. From each dataset, the centroids of each PASI thickness score were determined using k-means clustering. The centroid stability of every PASI thickness score in the complete dataset (see Table 4.8) and two subdatasets were then analyzed. The centroids of each score from the complete dataset and two subdatasets are shown in Table 4.8.

From Table 4.8 it can be seen that all scores except for score 4 have high centroid stability. Score 4 centroid stability is slightly affected due to the smaller number of data points. The k-means clustering needs a large input dataset to produce stable clustering because the centroid is the mean of the data points in the cluster and it is easily affected by noise and outliers. Statistically, a large number of data points means >30, hence the data points in each subdataset should be >30, giving a total complete dataset of >60. Score 4, which only has forty-five data points in the complete dataset, has the maximum centroid difference between the sets—0.012 mm. Nevertheless, this maximum centroid difference is still within the accepted error tolerance of 0.04 mm (10 times the PRIMOS 3D scanner's elevation resolution of 4 μm). The centroids are thus proven stable and the determined thickness range can be used as the reference to perform consistent PASI scoring.

4.4.2.3 Reliability of the Lesion Thickness Algorithm and PASI Scoring

The reliability of the lesion thickness algorithm and determined range of PASI thickness scoring was determined by analyzing the interrater agreement between thickness scores of two different 3D skin surface images acquired from one lesion. The interrater agreement value was calculated to determine if the PASI thickness scoring using the lesion thickness algorithm is reliable for repeatable assessment in clinical practice. The Cohen's kappa interrater agreement analysis was used to test scoring reliability.

The 3D skin surface images of 129 plaque psoriasis lesions were acquired twice with modified positions. The acquired 3D images were in-focus and on non-concave surfaces. The 3D images were then analyzed using the lesion thickness algorithm to determine lesion thicknesses. The thickness values were then scored based on the PASI thickness range listed in Table 4.7.

Of the 129 plaque psoriasis lesions, only 17 lesions were scored differently. All different scores were still within 1 score difference (e.g., score 1 and 2, score 2 and 3). The kappa interrater agreement value of the assessment was found to be 0.81, which means that the intrarater reliability of the lesion thickness algorithm and the determined range of PASI thickness scoring are outstanding. This outstanding agreement shows that the proposed measurement algorithm and scoring are reliable for use to monitor the severity level of lesion thickness as well as treatment efficacy in clinical practice.

4.5 Conclusions

Medical treatment of skin lesions includes creams, lotions, phototherapy, oral medications, and injection. To ensure treatment efficacy, dermatologists need to assess the severity of the skin lesion by observing the morphological properties of the lesion. Thickness is one of these properties, as well as area, distribution, color, inflammation, shape, and roughness. Periodic assessment of the severity level demonstrates the effectiveness of the treatment applied.

Dermatologists assess the thickness of the skin lesion using tactile inspection. They touch and feel the lesion, comparing it to the surrounding normal skin. Tactile inspection is performed several times and in several directions in order to provide a comprehensive assessment based on the dermatologist's knowledge and experience. This type of inspection does not result in a quantitative measure of lesion thickness and thus the outcome can vary between dermatologists (interrater variation) and between repetitive assessments by a single dermatologist (intrarater variation).

To overcome the subjectivity and variations in thickness assessments, an objective technique to provide a quantitative measure of lesion thickness has been developed. It has been reported that objective measures of skin and lesion thickness can be achieved using microscope, X-ray, and ultrasound technology. However, these methods measure the abnormal thickening of the epidermis layer caused by the skin lesion's occurrence instead of the lesion thickness above the normal skin.

In this chapter, an objective measurement method that adapts the dermatologist's tactile inspection, which can only perceive alteration of the skin lesion compared with the surrounding normal skin, was developed. The developed algorithm measures the thickness of the elevated lesion surface from the lesion base. It overcomes the problem posed by the surface curvature of human skin at different body locations.

In the lesion thickness assessment method, a skin surface image comprising the lesion and its surrounding normal skin is acquired using a 3D fringe projection scanner. For consistent quality in 3D image data acquisition, the settings of the 3D scanner and the ambient lighting are standardized. From the acquisition of the 3D skin surface image, the elevation of the scanned skin surface is obtained. The elevation of the lesion is measured from the base of the lesion to the top surface. However, the lesion base, which was formerly the normal skin, has to be determined first. The lesion base is assumed to have the same surface trend as the surrounding normal skin, therefore it is estimated from the surrounding normal skin using a 3D interpolation method. The normal skin is segmented from the lesion using a thresholding method. In the developed 3D automatic segmentation, the curved human skin surface is eliminated to produce a flat skin surface. Thus the elevation difference between the skin lesion and normal skin as well as its intensity difference on the corresponding 2D elevation map is raised, and this allows the thresholding method to be used in segmenting the skin lesion and normal skin. However, the automatic segmentation is limited to convex surfaces and cannot be applied to concave surfaces. In practice, this can be avoided by choosing reference lesions occurring on nonconcave skin surfaces.

The segmentation creates a mask $I_m(x, y)$ from which the coordinates of the normal skin (x_n, y_n) and lesion (x_l, y_l) are known. The coordinates are then used to find the respective elevation of normal skin (z_n) and the lesion (z_l) on the 3D skin surface image. Two surfaces are thus determined: the normal skin, $I_n(x, y, z)$, and the lesion, $I_l(x, y, z)$. The segmented normal skin is then best fitted using 3D polynomial surface fitting, resulting in an estimated surface, $I_e(x, y, z)$. The lesion base is the best-fitted surface on the lesion coordinates (x_l, y_l). The 3D skin surface image $I(x, y, z)$ is then subtracted from the estimated surface $I_e(x, y, z)$ only at the lesion coordinates (x_l, y_l) to determine the lesion elevations, which are then averaged to obtain the lesion thickness, t.

The method also determines the thickness ranges of the lesion for each severity level of the standard assessment tool. Plaque psoriasis, as one of the most common skin diseases, was chosen as the object of this study. In clinical practice, the dermatologist uses the PASI to assess the severity level of psoriasis in patients. A total of 713 3D skin surface images comprising the lesion and surrounding normal skin were acquired from male and female patients registered in the Dermatology Department, Hospital Kuala Lumpur. To acquire nonconcave 3D skin surface images on concave body surfaces such as the chest and waist, patients were asked to bend their body. A classification method was used to group the measured thickness into PASI thickness scores of 1 to 4, where a higher score shows a more marked elevation of the psoriasis lesion. The interrater agreement of the dermatologists' scores was found to be $\kappa = 0.52$, and thus the dermatologists' scores are not suitable as a priori knowledge for the classification. The unsupervised classification method k-means clustering was chosen instead. Using k-means for score classification, a thickness range of 0.053–0.177 mm is classified as score 1, 0.178–0.277 mm as score 2, 0.278–0.439 mm as score 3, and >0.440 mm as score 4.

Centroid stability analysis was used to test the performance of the k-means clustering in determining the thickness range of the PASI thickness score. From the result, the maximum centroid difference of 0.012 mm found in score 4 was still within the accepted error tolerance of 0.04 mm. Thus the determined thickness range can be used as the reference to perform consistent PASI thickness scoring. The reliability of the lesion thickness algorithm and the range of PASI thickness scoring were determined by analyzing the interrater agreement between thickness scores of two different 3D skin surface images acquired from one lesion. The 3D skin surface images of 129 plaque psoriasis lesions were acquired twice with a modified position. The thicknesses of the lesions were measured using the lesion thickness algorithm and the PASI thickness score was determined according to the clustered thickness range. The results show that only seventeen lesions were scored differently. The kappa interrater agreement value of the assessment was found to be 0.81, which means that the intrarater reliability of the lesion thickness algorithm and the range of PASI thickness scoring are outstanding. This outstanding agreement shows that the proposed measurement algorithm and scoring are reliable for use in monitoring the severity of lesion thickness as well as treatment efficacy in clinical practice.

In the application of the lesion thickness algorithm, there are two constraints for successful thickness measurements and PASI thickness scoring. The first constraint is that the size of the plaque psoriasis lesion should not be larger than the 4 cm × 3 cm measurement area of the PRIMOS 3D scanner since the presence of a minimum of 50% surrounding normal skin on the 3D skin surface image is required to construct an accurate lesion base. This limitation can be overcome by utilizing another type of 3D scanner that has a wider measurement area, but some adjustments to the algorithm will be needed. It should also be noted that other types of 3D scanners have a greater depth of field so accurate thickness measurements can be achieved for high-curvature skin surfaces.

The second constraint is that the lesion or surrounding normal skin cannot be covered by dense or thick hair since the hair adds nonskin elevations to the acquired skin surface profile. This limitation can be overcome by designing a method that can remove hairs during the 3D image processing. The developed lesion thickness algorithm can be extended to other aspects of elevation measurement, including applications in terrain measurement (geoinformatics). A 3D surface profile of the terrain—digital elevation model (DEM) or digital terrain model (DTM)—can be imaged using field surveying, photogrammetry, or cartographic digitization. The lesion thickness algorithm can be applied in both regular and irregular grids of the DEM and DTM.

4.A Appendix: Algorithm MATLAB code

```
% LESION THICKNESS ALGORITHM
% Input: Z (3D skin surface image)
% Output : thickf (lesion thickness)

% 1. 3D AUTOMATIC SEGMENTATION

% Calculating window size of average filter based on the size of image

= size(Z);
b = k+1;

if b< = 200
    n = b/8;
end;
if b>200 && b< = 250
    n = b/6;
end;
if b>250 && b< = 300
    n = b/4;
end;
if b>300 && b< = 500
    n = (b/4)+15;
end;
if b>500 && b< = 600
    n = b/3;
end;

n = round(n);
n = n + mod(n-1,2);

% Applying average filter to 3D skin surface image

h = fspecial('average',[n n]);
Zx = imfilter(Z,h,'replicate');

% Calculating vertical deviation

kur = (Z-Zx);

% Capping negative values to 0

for v = 1:k
  for w = 1:l
    if kur(v,w) < = 0
        kur2(v,w) = 0;
    else
        kur2(v,w) = kur(v,w);
      end;
    end;
  end;

% Applying median filter with 15x15 window size

m = 15;
kur3 = medfilt2(kur2,);
```

```
% Transforming surface to grayscale

a = min(min(kur3));
b = max(max(kur3));
kur4 = ((kur3-a)/(b-a)).*255;
pic = uint8(kur4);

% Segmenting the normal skin and lesion using thresholding method, >100 is
%considered as lesion

for i = 1:(k)
    for j = 1:(l)
        if pic(i,j)>100
            A(i,j) = 1;
        else
            A(i,j) = 0;
        end;
      end;
    end;

% Determining the designation of pixels to normal skin and lesion

[x0 y0] = find(A = =0);
[x1 y1] = find(A = =1);

for i = 1:length(x0)
    a = x0(i);
    b = y0(i);
    z0(i,1) = Z(a,b);
 end

for i = 1:length(x1)
    a = x1(i);
    b = y1(i);
    z1(i,1) = Z(a,b);
 end

% 2. 3D LESION BASE ESTIMATION
n = length(x0);
[sizex sizey] = size(Z);
Z = ones(sizex,sizey);

% 3D surface fitting using 3rd order of polynomial function

V(:,1)  = ones(n,1);
V(:,2)  = y0;
V(:,3)  = y0.^2;
V(:,4)  = y0.^3;
V(:,5)  = x0;
V(:,6)  = x0.*y0;
V(:,7)  = x0.*(y0.^2);
V(:,8)  = x0.^2;
V(:,9)  = (x0.^2).*y0;
V(:,10) = x0.^3;

a = pinv(V)*z0;

for i = 1:sizex
    for j = 1:sizey
        v1 = 1;
```

```
        v2 = j;
        v3 = j.^2;
        v4 = j.^3;
        v5 = i;
        v6 = i.*j;
        v7 = i.*(j.^2);
        v8 = i.^2;
        v9 = (i.^2).*j;
        v10 = i.^3;
        v = [v1 v2 v3 v4 v5 v6 v7 v8 v9 v10];

        Znew3(i,j) = v*a;
      end
    end

for i = 1:length(x0)
    a = x0(i);
    b = y0(i);
    z0new3(i,1) = Znew3(a,b);
end

for i = 1:length(x1)
    a = x1(i);
    b = y1(i);
    z1new3(i,1) = Znew3(a,b);
 end
% 3D surface fitting using 4th order of polynomial function

V(:,1) = ones(n,1);
V(:,2) = y;
V(:,3) = y.^2;
V(:,4) = y.^3;
V(:,5) = y.^4;
V(:,6) = x;
V(:,7) = x.*y;
V(:,8) = x.*(y.^2);
V(:,9) = x.*(y.^3);
V(:,10) = x.^2;
V(:,11) = (x.^2).*y;
V(:,12) = (x.^2).*(y.^2);
V(:,13) = (x.^3);
V(:,14) = (x.^3).*y;
V(:,15) = (x.^4);

a = pinv(V)*z;

for i = 1:sizex
    for j = 1:sizey
        v1 = 1;
        v2 = j;
        v3 = j.^2;
        v4 = j.^3;
        v5 = j.^4;
        v6 = i;
        v7 = i.*j;
        v8 = i.*(j.^2);
        v9 = i.*(j.^3);
        v10 = i.^2;
```

```
        v11 = (i.^2).*j;
        v12 = (i.^2).*(j.^2);
        v13 = (i.^3);
        v14 = (i.^3).*j;
        v15 = (i.^4);
        v = [v1 v2 v3 v4 v5 v6 v7 v8 v9 v10 v11 v12 v13 v14 v15];

        Znew4(i,j) = v*a;
      end
   end

for i = 1:length(x0)
   a = x0(i);
   b = y0(i);
   z0new4(i,1) = Znew4(a,b);
 end

for i = 1:length(x1)
   a = x1(i);
   b = y1(i);
   z1new4(i,1) = Znew4(a,b);
 end

% 3D surface fitting using 5th order of polynomial function

V(:,1)  = ones(n,1);
V(:,2)  = y;
V(:,3)  = y.^2;
V(:,4)  = y.^3;
V(:,5)  = y.^4;
V(:,6)  = y.^5;
V(:,7)  = x;
V(:,8)  = x.*y;
V(:,9)  = x.*(y.^2);
V(:,10) = x.*(y.^3);
V(:,11) = x.*(y.^4);
V(:,12) = x.^2;
V(:,13) = (x.^2).*y;
V(:,14) = (x.^2).*(y.^2);
V(:,15) = (x.^2).*(y.^3);
V(:,16) = (x.^3);
V(:,17) = (x.^3).*y;
V(:,18) = (x.^3).*(y.^2);
V(:,19) = (x.^4);
V(:,20) = (x.^4).*y;
V(:,21) = (x.^5);

a = pinv(V)*z;

[sizex sizey] = size(ZI);
Z = ones(sizex,sizey);

for i = 1:sizex
    for j = 1:sizey
        v1 = 1;
        v2 = j;
        v3 = j.^2;
        v4 = j.^3;
        v5 = j.^4;
        v6 = j.^5;
```

```
        v7 = i;
        v8 = i.*j;
        v9 = i.*(j.^2);
        v10 = i.*(j.^3);
        v11 = i.*(j.^4);
        v12 = i.^2;
        v13 = (i.^2).*j;
        v14 = (i.^2).*(j.^2);
        v15 = (i.^2).*(j.^3);
        v16 = (i.^3);
        v17 = (i.^3).*j;
        v18 = (i.^3).*(j.^2);
        v19 = (i.^4);
        v20 = (i.^4).*j;
        v21 = (i.^5);
        v = [v1 v2 v3 v4 v5 v6 v7 v8 v9 v10 v11 v12 v13 v14 v15 v16 v17 v18
        v19 v20 v21];

        Znew5(i,j) = v*a;
      end
    end

   for i = 1:length(x0)
      a = x0(i);
      b = y0(i);
      z0new5(i,1) = Znew5(a,b);
   end

for i = 1:length(x1)
     a = x1(i);
     b = y1(i);
     z1new5(i,1) = Znew5(a,b);
   end

   % Calculating fitting error and thickness of lesion from each polynomial
   order; negative elevation is capped to zero

% 3rd order
error3 = mean(mean(abs(z0-z0new3)));
elevation3 = z1-z1new3;
elevation3pos = [];
for i = 1:length(elevation3)
   if elevation3(i)>0
      elevation3posi = elevation3(i);
      elevation3pos = [elevation3pos;elevation3posi];
    end
  end
thick3 = mean(elevation3pos);

% 4th order
error4 = mean(mean(abs(z0-z0new4)));
elevation4 = z1-z1new4;
elevation4pos = [];
for i = 1:length(elevation4)
  if elevation4(i)>0
     elevation4posi = elevation4(i);
     elevation4pos = [elevation4pos;elevation4posi];
  end
end
thick4 = mean(elevation4pos);
```

```
% 5th order
error5 = mean(mean(abs(z0-z0new5)));
elevation5 = z1-z1new5;
elevation5pos = [];
for i = 1:length(elevation5)
   if elevation5(i)>0
       elevation5posi = elevation5 (i);
       elevation5pos = [elevation5pos;elevation5posi];
   end
end
thick5 = mean(elevation5pos);

error = [error3;error4;error5];
thick = [thick3;thick4;thick5];

% 3. DETERMINATION OF LESION SURFACE ELEVATION FOR THICKNESS

% Choosing thickness from fitted surface with minimum fitting error to
% segmented normal skin

errorf = min(error);
e = find(error = =errorf);
thickf = thick(max(e));
```

4.B Appendix: Area-Based Surface Fitting

In area-based surface fitting, the fitting process uses all values of all points in the surface area. Based on the fitting purpose, the surface fitting is divided into two types, best fitting and exact fitting. A best-fitting method is applied to the surface if the purpose of the fitting is to obtain a fitted surface that follows its trend. An exact fitting is applied to the surface if the purpose of the fitting is to obtain a fitted surface that exactly fits the original surface.

Best Fitting

Best fitting is suited for rough surfaces because it shows the curvature of the surface. A least-squares method using polynomial functions is the simplest and most common method to perform best fitting [42]. The orders of the polynomial function show the number of surface oscillations used to fit a surface, where a higher order means more surface oscillation. The second, third, and fourth orders create best fitting with one, two, and three surface oscillations, respectively. Higher-order polynomials can be applied to get more surface oscillation, resulting in greater accuracy of the fitting. However, applying higher orders of polynomial functions may overfit the surface, resulting in oscillations on the border of the surface, or Runge's phenomenon [43].

The polynomial function for area-based surface fitting using the best-fitting method is the same function used in point-based interpolation as listed in Table 4.B.1. Instead of performing the fitting using neighbor points, as in the point-based interpolation, this best-fitting method employs all points on the surface.

There are two steps to perform the best fitting on a 3D surface using a polynomial function. First, a polynomial coefficient is determined using a matrix of the coordinates $[X_s, Y_s]$ and its respective elevation value $[Z_s]$ (see equation (4.B.1)).

$$[a] = [X_s, Y_s]^{-1}[Z_s] \tag{4.B.1}$$

After obtaining the matrix of the polynomial coefficient, $[a]$, a fitted surface with elevation values $[Z_e]$ on its respective coordinates $[X_e, Y_e]$ is determined using equation (4.B.2).

$$[Z_s] = [X_e, Y_e][a] \tag{4.B.2}$$

Exact Fitting

A higher-order polynomial function can be used to obtain a fitted surface that exactly fits the original surface. However, the Runge phenomena may occur. A bicubic spline interpolation is then introduced as the most common method for performing an exact fitting that requires a lower degree of polynomial function. The bicubic spline interpolation is shown in equation (4.B.3) [44].

$$z = f(x, y) = \sum_{j=0}^{3} \sum_{i=0}^{3} a_{i,j} x^i y^j \tag{4.B.3}$$

$$= a_{00} + a_{10}x + a_{20}x^2 + a_{30}x^3 + a_{01}y + a_{11}xy + a_{21}x^2y$$

$$+ a_{31}x^3y + a_{02}y^2 + a_{12}xy^2 + a_{22}x^2y^2 + a_{32}x^3y^2 a_{03}y^3$$

$$+ a_{13}xy^3 + a_{23}x^2y^3 + a_{33}x^3y$$

TABLE 4.B.1

Polynomial Function [44]

Order	Descriptive Term	Function
Zero	Planar	$Z = f(x, y) = a_0$
First	Linear	$Z = f(x, y) = a_0 + a_1x + a_2y$
Second	Quadratic	$Z = f(x, y) = a_0 + a_1x + a_2y + a_3x^2 + a_4y^2 + a_5xy$
Third	Cubic	$Z = f(x, y) = a_0 + a_1x + a_2y + a_3x^2 + a_4y^2 + a_5xy + a_6x^3 + a_7y^3 + a_8x^2y + a_9xy^2$
Fourth	Quartic	$Z = f(x, y) = a_0 + a_1x + a_2y + a_3x^2 + a_4y^2 + a_5xy + a_6x^3 + a_7y^3 + a_8x^2y + a_9xy^2 + a_{10}x^4 + a_{11}y^4 + a_{12}x^3y + a_{13}x^2y^2 + a_{14}xy^3$
Fifth	Quintic	$Z = f(x, y) = a_0 + a_1x + a_2y + a_3x^2 + a_4y^2 + a_5xy + a_6x^3 + a_7y^3 + a_8x^2y + a_9xy^2 + a_{10}x^4 + a_{11}y^4 + a_{12}x^3y + a_{13}x^2y^2 + a_{14}xy^3 + a_{15}x^5 + a_{16}x^5 + a_{17}x^4y + a_{18}x^3y^2 + a_{19}x^2y^3 + a_{20}xy^4$

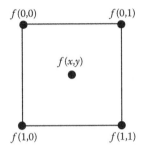

FIGURE 4.B1
Bicubic spline interpolation using four neighbor points [45].

The bicubic spline interpolation used to determine the value of $f(x, y)$ is illustrated in Figure 4.B.1. The 16 coefficients on equation (4.B.2) are determined on each piecewise surface using the 16 equations (equations 4.B.4–4.B.19) of the function values; derivatives along the x direction, derivatives along the y direction, and cross-derivatives along the x–y direction form four neighbor points [45]. Function values:

$$f(0,0) = a_{00} \tag{4.B.4}$$

$$f(1,0) = a_{00} + a_{10} + a_{20} + a_{30} \tag{4.B.5}$$

$$f(0,1) = a_{00} + a_{01} + a_{02} + a_{03} \tag{4.B.6}$$

$$f(1,1) = \sum_{j=0}^{3}\sum_{i=0}^{3} a_{ij} = a_{00} + a_{01} + a_{02} + a_{03} + a_{10} + a_{11} + a_{12} + a_{13} + a_{20} + a_{21} + a_{22} + a_{23}$$

$$+ a_{30} + a_{31} + a_{32} + a_{33} \tag{4.B.7}$$

Derivatives along the x direction:

$$\frac{\partial f(0,0)}{\partial x} = a_{10} \tag{4.B.8}$$

$$\frac{\partial f(1,0)}{\partial x} = a_{10} + a_{20} + a_{30} \tag{4.B.9}$$

$$\frac{\partial f(0,1)}{\partial x} = a_{10} + 2a_{20} + 3a_{30} \tag{4.B.10}$$

$$\frac{\partial f(1,1)}{\partial x} = \sum_{j=0}^{3}\sum_{i=0}^{3} a_{ij}^{i} \tag{4.B.11}$$

Derivatives along the y direction:

$$\frac{\partial f(0,0)}{\partial y} = a_{01} \tag{4.B.12}$$

$$\frac{\partial f(1,0)}{\partial y} = a_{01} + a_{02} + a_{03} \tag{4.B.13}$$

$$\frac{\partial f(0,1)}{\partial y} = a_{01} + 2a_{02} + 3a_{03} \tag{4.B.14}$$

$$\frac{\partial f(1,1)}{\partial y} = \sum_{j=0}^{3}\sum_{i=0}^{3} a_{ij}^{j} \tag{4.B.15}$$

Cross-derivatives along the x–y direction:

$$\frac{\partial^2 f(0,0)}{\partial x \partial y} = a_{11} \tag{4.B.16}$$

$$\frac{\partial^2 f(1,0)}{\partial x \partial y} = a_{11} + 2a_{21} + 3a_{31} \tag{4.B.17}$$

$$\frac{\partial^2 f(0,1)}{\partial x \partial y} = a_{11} + 2a_{12} + 3a_{13} \tag{4.B.18}$$

$$\frac{\partial^2 f(1,1)}{\partial x \partial y} = \sum_{j=0}^{3}\sum_{i=0}^{3} a_{ij}^{ij} \tag{4.B.19}$$

4.C Appendix: Kappa Interrater Agreement Analysis

The per-pixel segmentation performance in the medical imaging application is determined using interrater agreement analysis, where the designation of each pixel in the segmentation result is compared with the pixel designation in the reference [46, 47]. Kappa is a statistical measure used in the interrater agreement analysis for the categorical data [48]. It can determine the agreement between two or multiple raters and for two or multiple categories. To determine the interrater agreement between two raters with two or multiple categories, Cohen's kappa is used. Given two 2D masking of reference A and segmentation result B with n number of pixels that is designated into q number of categories, the overall agreement probability p_a is the proportion of pixels that both 2D maskings designated into the same category. It is common to all categories and given by equation 4.C.1 [49]:

$$p_a = \sum_{k=1}^{q} n_{kk}/n \qquad (4.C.1)$$

where n_{kk} is the number of pixels that is designated into the same category. Let n_{Ak} and n_{Bk} be the number of pixels that are designated into each category by set A and B, respectively. Then the chance agreement probability $p_{e|k}$ that is specific to each category is calculated using equation (4.C.2):

$$p_{e|k} = \sum_{k=1}^{q} n_{Ak} n_{Bk}/n^2 \qquad (4.C.2)$$

Cohen's kappa, $\hat{\gamma}_k$, is then calculated using equation (4.C.3):

$$\hat{\gamma}_k = (p_a - p_{e|k})/(1 - p_{e|k}) \qquad (4.C.3)$$

The kappa statistics value is from 0 to 1 and is interpreted as the agreement strength between both 2D maskings, as shown in Table 4.C.1 [50].

A kappa value of 0.60–0.79 is considered to have substantial agreement between both 2D maskings and a kappa value of 0.80–1.00 is considered to have outstanding agreement beyond chance.

TABLE 4.C.1

Agreement Strength of the Kappa Statistics [50]

Kappa	Agreement Strength
<0.00	Poor
0.00–0.19	Slight
0.21–0.39	Fair
0.40–0.59	Moderate
0.60–0.79	Substantial
0.80–1.00	Outstanding

References

1. Zaidi Z, Lanigan SW. *Dermatology in clinical practice*. London: Springer, 2010.
2. Shimizu H. *Shimizu's textbook of dermatology*. Hokkaido: Hokkaido University Press, 2007.
3. Baird KA. A simplified classification of skin diseases. *Canadian Medical Association Journal* 1943;49:200–204.
4. Lawton S. Assessing the patient with a skin condition. *Practice Nurse* 2005; 30:43–48.
5. Papadopoulos L, Walker C. *Understanding skin problem: acne eczema, psoriasis and related conditions*. Sussex, UK: John Wiley & Sons, 2003.
6. Miyamoto K, Kaczvinsky J, Robinson L, Deng G. Measurement with a new in-vivo skin topographical method of facial wrinkle improvement by skin moisturizers formulated with anti-aging ingredients. Paper presented at the American Association for the Advancement of Science (AAAS) Conference Meeting, 2009.
7. Mitchell T, Penzer R. *Psoriasis at your fingertips: the comprehensive and medically accurate manual on managing psoriasis*. London: Class Publishing, 2000.
8. Fry L. *An atlas of psoriasis*, 2nd ed. London: Taylor & Francis, 2004.
9. Dermatological Society of Malaysia. http://www.dermatology.org.my
10. Sinniah B, Saraswathy Devi S, Prashant BS. Epidemiology of psoriasis in Malaysia: a hospital based study. *Medical Journal Malaysia* 2010;65:112–114.
11. Lisi P. Differential diagnosis of psoriasis. *Reumatismo* 2007;59:56–60.
12. Lowe NJ. *Psoriasis: patient's guide*. London: Martin Dunitz, 2003.
13. van der Kerkhof M. *Textbook of psoriasis*. Oxford: Blackwell, 2003.
14. Alper M, Kavak A, Parlak AH, Demirci R, Belenli I. Measurement of epidermal thickness in a patient with psoriasis by computer-supported image analysis. *Brazilian Journal of Medical and Biological Research* 2004;37:111–117.
15. Uzun I, Akyildiz E, Inanici MA. Histopathological differentiation of skin lesion caused by electrocution, flame burns and abrasion. *Forensic Science International* 2008;178:157–161.
16. Olympus BX-50. http://www.olympus-global.com/cn/corc/history/micro/uis.cfm
17. Braun RP, Kaya G, Masouyé I, Krischer J, Saurat JH. Histopathological correlation in dermoscopy: a micropunch technique. *Archive of Dermatology* 2003;139:349–351.
18. DL3. http://www.dermlite.com/dl3.html
19. De Rosa G, Mignogna C. The histopatology of psoriasis. *Reumatismo* 2007;59:46–49.
20. Marks R, Dykes PJ, Roberts E. The measurement of corticosteroid induced dermal atrophy by a radiological method. *Archives of Dermatological Research* 1975;253:93–96.
21. Black MM. A modified radiographic method for measuring skin thickness. *British Journal of Dermatology* 1969;81:661–666.
22. Alexander H, Miller DL. Determining skin thickness with pulsed ultrasound. *Journal of Investigative Dermatology* 1979;27:17–19.
23. Pellacani G, Seidenari S. Preoperative melanoma thickness determination by 20-MHz sonography and digital videomicroscopy in combination. *Archives of Dermatology* 2003;139:293–298.

24. Gupta AK, Turnbull DH, Harasiewicz KA, Shum DT, Watteel GN, Foster FS, Sauder DN. The use of high-frequency ultrasound as a method of assessing the severity of a plaque psoriasis. *Archives of Dermatology* 1996;132:658–662.

25. Vaillant L, Berson M, Machet L, Callens A, Pourcelot L, Lorette G. Ultrasound imaging of psoriatic skin: a noninvasive technique to evaluate treatment of psoriasis. *International Journal of Dermatology* 1994;33:786–790.

26. Serup J. Non-invasive quantification of psoriasis plaque measurement of skin thickness with 15 mHz pulsed ultrasound. *Clinical and Experimental Dermatology* 1984;9:502–508.

27. Vilana R, Puig S, Sanchez M, Squarcia M, Lopez A, Castel T, Malvehy J. Preoperative assessment of cutaneous melanoma thickness using 10-MHz sonography. *American Journal of Roentgenology* 2009;193:639–643.

28. Bekerecioğlu M, Arslan H, Uğrras S, Karakök M, Akpolat N. Comparison of thickness measurement in cutaneous and subcutaneous lesions preoperatively (by ultrasonography) and postoperatively (by ruler). *European Journal of Plastic Surgery* 1998;21:236–237.

29. Ultrasound virtual biopsy. http://www.a1med.net/ultrasound_virtual_biopsy.html

30. Ihtatho D. Objective assessment of area and erythema of psoriasis lesion using digital imaging and colourimetry. Master's thesis, Electrical and Electronics Engineering Department, University Technology Petronas, Perak, Malaysia, 2008.

31. Gomez DD, Carstensen JM, Ersbøll, Skov L, Bang B. Building an image-based system to automatically score psoriasis. In Bigun J, Gustavsson T, eds. *Image analysis: 13th Scandinavian Conference Proceedings*. Berlin: Springer, 2003:557–564.

32. Ramsay B, Lawrence CM. Measurement of involved surface area in patients with psoriasis. *British Journal of Dermatology* 1991;124:565–570.

33. Jacobi U, Chen M, Frankowski G, Sinkgraven R, Hund M, Rzany B, Sterry W, Lademann J. In vivo determination of skin surface topography using an optical 3D device. *Skin Research and Technology* 2004;10:207–214.

34. Smalls LK, Lee CY, Whitestone J, Kitzmiller WJ, Wickett RR, Visscher MO. Quantitative model of cellulite: three dimensional skin surface topography, biophysical characterization and relationship to human perception. *Journal of Cosmetic Science* 2005;56:105–120.

35. Cula OG, Dana KJ, Murphy FP, Rao BK. Bidirectional imaging and modeling of skin texture. *IEEE Transactions on Biomedical Engineering* 2004;51:2148–2159.

36. GFMesstechnik. *PRIMOS users guide*. Berlin: GFMesstechnik, 2007.

37. Jones BF, Plassmann P. An instrument to measure the dimension of skin wounds. *IEEE Transactions on Biomedical Engineering* 1995;42:464–470.

38. Berry JK. *Map analysis: understanding spatial patterns and relationships*. San Francisco: GeoTech Media, 2007.

39. Luntama JP, Koponen S, Hallikainen M. Analysis of sea ice thickness and mass estimation with a spaceborne laser altimeter. *IEEE International Geoscience and Remote Sensing Symposium Proceedings*. Washington, DC: IEEE, 1997:1314–1316.

40. Rosenthal MR. Effective thickness of paper: appraisal and further development. Research paper FPL 287. Madison, WI: USDA Forest Products Laboratory, 1977.

41. Chang CC, Gangaram HB, Hussein SH. Malaysian Psoriasis Registry—preliminary report of a pilot study using newly revised registry form. *Medical Journal Malaysia* 2008;63:68–71.

42. Lancaster P, Salkauskas K. *Curve and surface fitting: an introduction*. London: Academic Press, 1986.

43. Chen Y. High-order polynomial interpolation based on the interpolation center's neighborhood. *World Congress on Software Engineering Proceedings*. Washington, DC: IEEE, 2009:345–348.

44. Li Z, Zhu C, Gold C. *Digital terrain modeling: principles and methodology*. New York: CRC Press, 2005.

45. Giassa M. Image processing—bicubic spline interpolation. http://www.giassa.net/

46. Colliot O, Mansi T, Bernasconi N, Naessens V, Klironomos D, Bernasconi A. A level set driven by MR features of focal cortical dysplasia for lesion segmentation. *Medical Image Understanding and Analysis 2005. Proceedings of the 9th annual conference*. Bristol: University of Bristol, 2005:239–242.

47. Oishi K, Faria A, Jiang H, Li X, Akhter K, Zhang J, Hsu JT, Miller MI, van Zijl PCM, Albert M, Lyketsos CG, Woods R, Toga AW, Pike GB, Rosa-Neto P, Evans A, Mazziotta J, Mori S. Atlas-based whole brain white matter analysis using large deformation diffeomorphic metric mapping: application to normal elderly and Alzheimer's disease participants. *Neuroimage* 2009;46:486–499.

48. Gwet KL. *Handbook of inter-rater reliability*, 2nd ed. Gaithersburg, MD: Advanced Analytics, 2010.

49. Gwet KL. Computing inter-rater reliability and its variance in the presence of high agreement. *British Journal of Mathematical and Statistical Psychology* 2008;61:29–48.

50. Landis JR, Koch GG. The measurement of observer agreement for categorical data. *Biometrics* 1977;33:159–174.

51. H. Fitriyah, "Thickness Assessment of Skin Lesion Using 3D Skin Surface Imaging," Masters of Science Thesis, Electrical and Electronics Engineering Department University Teknologi PETRONAS, Perak, Malaysia, 2012.

52 Ahmad Fadzil M Hani, H. Fitriyah, E. Prakasa, et. al "In vivo 3D thickness measurement of skin lesion, in *IEEE EMBS Conference on Biomedical Engineering and Sciences*, 2010, pp. 155–160.

5

Analysis of Skin Pigmentation

Ahmad Fadzil Mohamad Hani, Hermawan Nugroho and
Norashikin Shamsudin

CONTENTS

5.1 Pigmentary Skin Disorder

Vitiligo is a pigmentary skin disorder resulting from abnormal melanin pro-duction. This is due to the destruction of epidermal melanocytes. Physically, vitiligo lesion areas appear as paler skin tone compared with the surround-ing skin or they may be completely white [1]. Figure 5.1 shows a patient with vitiligo lesions. The prevalence of vitiligo worldwide varies, ranging from 0.1% to 2%. The prevalence of vitiligo is not related to any skin types, sex, or socioeconomic status [2]. Vitiligo is unpredictable and may remain stable for years before worsening. The disease is most disfiguring in dark-skinned racial and ethnic groups where the contrast between the depigmented and

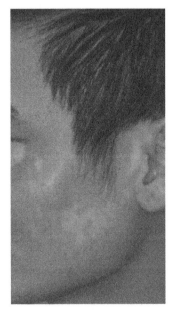

FIGURE 5.1
White skin patches on a vitiligo patient.

healthy skin is more noticeable. It has been reported that patients with vitiligo have an increased risk of autoimmune diseases such as thyroid disease (Hashimoto's thyroiditis and Grave's disease) and Addison's disease [2, 3].

5.1.1 Clinical Features of Vitiligo

Vitiligo is seen as acquired white or hypopigmented maculae or patches, as shown in Figures 5.2–5.6. The disease is categorized according to the distribution and the extent of involvement of depigmentation. However, clinical assessments may vary and can be categorized as non-segmental vitiligo (generalized, acral or acrofacial, localized or focal) and segmental vitiligo.

5.1.1.1 Generalized Vitiligo

Figure 5.2 depicts the generalized type, which is the most common pattern, with bilateral, symmetric depigmentation of the face (typically the periorificial areas), torso, neck, extensor surfaces, or bony prominences of the hands, wrists, and legs.

5.1.1.2 Acrofacial Vitiligo

Typically acrofacial vitiligo is limited to the distal digits and periorificial facial areas, the latter in a circumferential pattern (Figure 5.3).

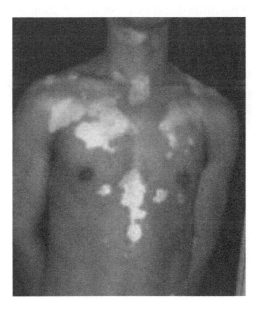

FIGURE 5.2
Generalized vitiligo (courtesy of Hospital Kuala Lumpur).

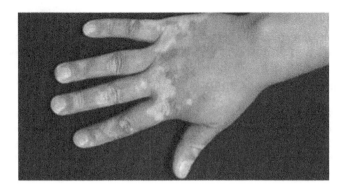

FIGURE 5.3
Acrofacial vitiligo on the distal digits (courtesy of Hospital Kuala Lumpur).

5.1.1.3 Segmental Vitiligo

Segmental vitiligo, as shown in Figure 5.4, is the least common pattern and occurs in a unilateral, dermatomal, or quasi-dermatomal distribution, often following the distribution of the trigeminal nerve. It is known for its early onset and rapid initial growth with nonprogression within two years.

5.1.1.4 Focalized Vitiligo

Focalized vitiligo has a limited and localized distribution, as seen in Figure 5.5. However, it may develop into generalized vitiligo or follow a stable course.

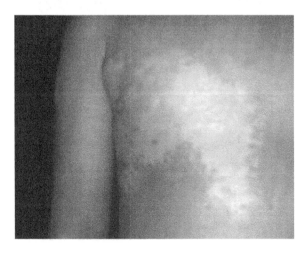

FIGURE 5.4
Segmental vitiligo (courtesy of Hospital Kuala Lumpur).

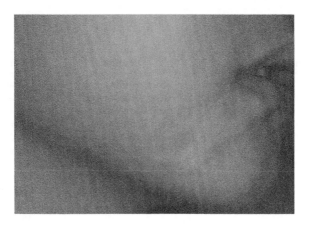

FIGURE 5.5
Focal vitiligo (courtesy of Hospital Kuala Lumpur).

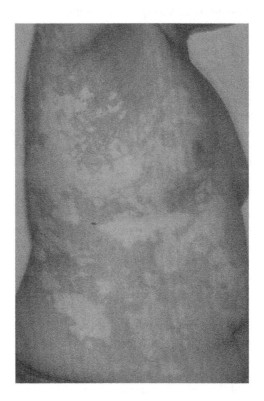

FIGURE 5.6
Universal vitiligo (courtesy of Hospital Kuala Lumpur).

TABLE 5.1

Vitiligo Treatment

Medical Treatment	Surgical	UVB or laser therapy
1. Topical and systemic corticosteroids	1. Mini grafting	1. Psoralen with exposure to ultraviolet A radiation therapy (PUVA)
2. Tacrolimus ointment	2. Transplantation of cultured melanocytes	
3. Heliotherapy	3. Transplantation of noncultured melanocytes	
4. Depigmentation therapy with monobenzylether of hydroquinone		2. Narrow Band UVB therapy
5. Tacrolimus ointment		3. 380 nm laser
		4. Depigmentation with Q-switched ruby laser

5.1.1.5 Universal Vitiligo

Universal vitiligo implies loss of pigment melanin over the entire body surface area, as seen in Figure 5.6.

5.1.2 Efficacy Assessment of Vitiligo Treatment

Table 5.1 lists various treatments under three categories—medical, surgical, and ultraviolet B (UVB) or laser therapy—that are readily available for vitiligo.

Dermatologists visually assess vitiligo's therapeutic response to treatment based on the degree of skin repigmentation within affected areas over time. As the process for repigmentation is slow, the time needed for repigmentation to be discerned is normally longer than six months. The evaluation of repigmented areas is also largely dependent on human visual perception and judgment. Currently there is no objective way to measure and quantify the repigmentation response.

5.1.3 Digital Image Analysis of Vitiligo

Skin is considered to be a layered construction of epidermis, dermis, and hypodermis. All possible colors occurring within normal human skin could be analytically modeled by exploiting the physics of optics related to the optical interface between these layers. In other words, skin appearance and color are due to the combination of skin histological parameters interacting with ambient light. However, in digital imaging, color perception is created by combining three different spectral bands: red, green, and blue (RGB). Moreover, the image formation process within a digital camera is a process involving the spectrum of the incoming light, the spectral characteristic of the camera's charged-coupled device (CCD), and the spectral transmittance of the Bayer filter. The incoming light is a result of the light source properties and the properties of the reflecting material—in this case, the human skin.

In the past, digital image processing has been limited to digital image acquisition, enhancement, and image retrieval. Due to the availability of higher-resolution digital cameras and powerful personal computers, digital image analysis techniques are now being applied for medical applications. In clinical settings, the analysis of repigmentation of vitiligo areas is a complicated task, particularly because of the variability of the images in terms of resolution (scale), ambient lighting, and angle of observation. In addition, the size of the repigmented areas can be very small and difficult to discern visually. Digital image analysis can be applied to overcome these challenges faced by dermatologists.

5.2 Objective Assessment of Repigmentation

5.2.1 Current Assessment

Generally there are two objectives in vitiligo treatment: to stop the progression of the disease and to repigment the lesions [4–6]. To monitor the effectiveness of treatments, dermatologists need to examine the lesion directly, or indirectly using digital photos. These digital skin images are visually analyzed for purpose of diagnosis by dermatologists. At present, the disease is analyzed by comparing the patient's images before and after treatment. The images are studied in order to assess the therapeutic response of the treatment. Dermatologists have to be trained to make accurate assessments, and this process requires a high degree of skill and experience. Nevertheless, the technique is still subjective, and it is common to have different results due to the varying degrees of experience of dermatologists.

At present, the Physician's Global Assessment (PGA) scale is used by dermatologists for vitiligo assessment. The scale represents the degree of repigmentation within lesions over time. However, most of the studies on vitiligo treatments vary in duration and the number of PGA scale points used. The degree of repigmentation, which defines the success of the treatment, is generally set around 50% to 75% repigmentation based on the global impression of the dermatologist to the overall therapeutic responses of treatment [7–9]. It is difficult to compare treatment outcomes given the differences in the PGA scales used to assess repigmentation. Table 5.2 shows a general PGA scale that can be used for vitiligo assessment.

TABLE 5.2

Physician's Global Assessment scale

Repigmentation	Scale
0%–25%	Mild
26%–50%	Moderate
51%–75%	Good
76%–100%	Excellent to complete

The evaluation of the treatment to produce the score is primarily based on the visual appearance of the lesion and the dermatologist's judgment. The therapeutic response of vitiligo treatment is typically very slow. The change can be so subtle that the dermatologist may not be able to detect it. Consequently the disease is observed over long periods, typically every six months [10, 11].

It is also known that patients respond differently to vitiligo treatments. The therapeutic response of a particular treatment may be very different in different patients. It is therefore useful for dermatologists to know the efficacy of a particular treatment earlier, in order that treatment can be adjusted. Currently dermatologists observe patient's vitiligo skin areas with the help of digital imaging. They compare features of the vitiligo lesions before and after treatment. However, since there are no fixed image acquisition procedures, it is found that the vitiligo images sometimes are not well illuminated and, in most cases, calibrations of scale in the images are not performed. As a result, it is hard to determine the actual area and subsequent changes in size of the vitiligo lesions and repigmentation areas due to treatment. The current scoring protocol used to evaluate treatment outcomes is largely arbitrary and highly subjective. Studies have shown that the assessment method results in inter- and intraobserver variations [12]. There is no known validated quantitative tool for characterizing vitiligo lesions parametrically [13].

For these reasons, we felt it was necessary to develop a qualitative tool that is highly sensitive to assist dermatologists in objectively monitoring vitiligo treatment efficacy. The system should be able to analyze, determine, and quantify vitiligo skin and repigmentation areas efficiently and reliably. More importantly, it should provide objective efficacy assessment to assist dermatologists in making accurate diagnoses and assessing therapeutic response. The system should provide dermatologists with an objective tool to determine repigmentation areas.

5.2.2 Development of Objective Assessments

During vitiligo treatment, skin images are obtained using a digital camera. The skin images produced by the camera are presented as arrays of pixels having discrete intensity values. In signal processing, digital skin images are seen as two-dimensional images that contain information about the skin. Using computing techniques, skin images can be analyzed and used as tools for assisting dermatologists. Moreover, with the decrease in cost and increase in computation power of personal computers, we now have the ability to develop a sophisticated and cost-effective computer-based image analysis system. Computer-aided analysis of digital skin images offers quantitative and repeatable measurements, reducing the subjectivity of the diagnosis. In addition, it also has the potential to enable dermatologists to monitor vitiligo on a shorter time cycle.

Research work related to image processing analysis of skin and vitiligo lesions can be categorized into two main approaches. The first approach employs image processing techniques to identify and determine the areas of the skin lesions. The second approach uses techniques to extract of skin pigment (chromophores related to pigmentation) parameters from the skin image.

5.2.3 Image Processing Techniques in the Determination of Skin Lesion Areas

Image processing techniques can be used to identify and determine the areas of the lesions. Several techniques have been proposed: color space transformation, statistical modeling, and neural networks.

5.2.3.1 Color Space Transformation

Principal component transformation was proposed by Fischer to segment pigmented skin lesions [14]. Principal component transformation has been used for segmentation of skin tumors [15, 16], resulting in high detection rates. However, the method is inaccurate for vitiligo lesion detection. An image processing technique for detecting skin color areas in hue, saturation, and value (HSV) color space was proposed by Cho et al. [17]. The method assumes that the skin color of subjects is similar. Problems arise with subjects of different skin photo types (SPTs). A skin segmentation method in tint, saturation, and luminance (TSL) color space was developed by Tomaz et al. [18]. Digital skin RGB images are first transformed into TSL color space. The skin pixels lie on a particular ellipse in the TSL color space. The overall performance of this method is not very good compared with other methods. van Geel et al. [19] used the region growing technique in the CIELAB color space to segment vitiligo areas. It was reported that the performance of this digital image analysis technique is not significantly different from the measurement by doctors. The error increases with an increase in lesion area. A skin color segmentation technique in the YCgCr color space was proposed by de Dios and Garcia [20]. The skin color is easily segmented into the separated luminance and chrominance components. However, this method does not take into account human pigments and hence is not accurate.

5.2.3.2 Statistical Model and Simulation

A segmentation method for pigmented skin was proposed by Schmid based on two-dimensional (2D) histogram analysis and the fuzzy c-means (FCM) clustering technique [14]. FCM clustering has proven to be a robust clustering technique. The drawback of the method is that it needs to have prior knowledge of the pigmented lesion.

A statistical approach has also been applied by Jones and Rehg [21]. The statistical models were built using a general histogram density using a large image database taken from the Internet. The method is able to detect 80%

of skin pixels with a false-positive rate of 8.5%. The method requires a huge amount of training data and it cannot be used for pigmented skin. In another work based on statistical modeling by Caetano and Barone, skin colors are modeled using a mixture of Gaussian distributions [22]. The parameters of the model are derived iteratively using the expectation-maximization (EM) algorithm. The drawback of the method is that it only works to discern color combinations from two ethnic origins, Caucasian and African. Another EM algorithm approach was proposed by Zhu et al. in which the image is modeled by five Gaussian kernels [23]. To distinguish the skin component from other Gaussian components, the authors heuristically fix the parameter of the Gaussian kernel related to skin color. The drawback is that the method is not adaptive for detecting skin colors from different individuals. The method also requires a huge amount of training data to achieve a good detection rate.

A different approach using Gaussian kernels was proposed by Chang et al. [24]. In this work, a single Gaussian model is employed to fit the skin color distribution and FCM is employed to segment the skin. The method requires a huge number of different skin colors to achieve a high detection rate. A combination of wavelet and Gaussian modeling was developed by Kim and Kim [25]. Wavelet transformation was used to obtain a global view of an image by examining it at various resolution levels. The skin color was then segmented using a Gaussian model. The method does not take into account human pigment and hence it is not accurate. Skin segmentation classifiers based on Bayesian decision rules was proposed by Phung and Bouzerdoum [26]. Initially samples of skin pixels and nonskin pixels are collected and then probability distribution functions of skin pixels based on Bayesian rules are developed. The method depends on the data collection and requires a huge amount of training data to achieve a good detection rate.

5.2.3.3 Neural Networks

A neural network model for skin was developed by Seow and Asari [27]. It was observed that the skin colors form a nonlinear pipeline in the RGB space. A learning algorithm for a recurrent neutral network was proposed. Another work that used neural networks was developed by Mostafa and Abdelazeem [28]. The developed neural network uses normalized RGB color space as the input. The drawback of neural network–based segmentation techniques is that they need huge and good training images to perform segmentation on different skin colors.

Generally, segmentation based on image processing is not very accurate because the segmentation outcome is not necessarily lesion or skin. To be accurate, the method should take into account the skin and lesion makeup (i.e., skin chromophores and pigment concentrations).

5.2.3.4 Techniques in Extraction of Skin Pigments (Chromophores)

Takiwaki et al. [29] developed a formula to measure pigment melanin based on the reflectance values of the red channel in an RGB image. The formula is believed to be equivalent to the melanin index (MI) of narrowband spectrometers. Erythema and pigment melanin in port wine stain lesions were evaluated by Kelly et al. [30] and Jung et al. [31]. It is assumed that in the CIELAB color space, the L^* value represents the inverse of the melanin distribution map and a^* represents the erythema distribution map. Both methods simplify the interaction of light and skin and thus are not accurate. Since 1997, Cotton et al. have developed an optics-based skin model based on the CIE LMS color space for extracting histological skin parameters [32–35]. CIE LMS is a color space represented by the response of the three types of cones of the human eye, named after their sensitivity at long, medium, and short wavelengths. In another work by Tsumura et al., the spatial distribution of melanin and hemoglobin in skin was separated by employing linear independent component analysis (ICA) of a skin color image [36, 37]. The method assumes that the spatial variations of skin image color are caused by two skin pigments (melanin and hemoglobin) and their quantities are mutually independent. Segmentation based on ICA should be sufficiently accurate because the outcome of ICA enables us to determine the pigment melanin and hemoglobin of skin areas in the image.

Therefore, in our developed method, the ICA approach is employed. The objective of the method is to determine the repigmentation areas of vitiligo lesions over a shorter time period. The developed method consists of a variety of image processing techniques, including image rotation, geometric transformation, segmentation, and a variety of signal processing techniques such as ICA.

5.3 Objective Assessment of Vitiligo Using VT-Scan

5.3.1. Introduction to VT-Scan

A digital image analysis software called VT-Scan (Vitiligo Scan) for fast and objective assessment of therapeutic response to vitiligo treatments (treatment efficacy) was developed by researchers from the Centre for Intelligent Signal and Imaging Research (CISIR), Universiti Teknologi PETRONAS, Perak, Malaysia.

VT-Scan works on the principle of the skin color model, introduced by Tsumura et al. [36], that describes skin color as being spatially distributed according to melanin and hemoglobin pigments. The model enables effective principal component analysis (PCA) of digital RGB skin images, resulting in two principal components. This is followed by ICA to extract the independent components representing melanin and hemoglobin pigments. The resulting

FIGURE 5.7
VT-Scan software interface and system.

melanin-based image is then used to determine the melanin pigment composition of the skin. This will enable dermatologists to determine vitiligo and repigmentation areas accurately, allowing the assessment to be conducted within a shorter duration of six weeks compared with the typical three–six months.

Figure 5.7 shows the VT-Scan software interface and system consisting of a camera and computer. The following sections discuss the details of the skin color model, pigment analysis, segmentation, and measurements used in the VT-Scan algorithm.

5.3.2 Skin Color Model

5.3.2.1 Interaction of Light and Skin

When light encounters skin, a portion of the light is reflected due to the difference of the reaction index between the air and skin surface. This reflectance, known as surface reflectance, makes up 4% to 7% of the entire spectrum of 250–3000 nm [38]. Most of the light penetrates the skin, travelling along complex paths, and exits the skin or is absorbed by skin chromophores. A chromophore is the part of a molecule responsible for color by absorbing the electromagnetic energy of the light. In skin, the main chromophores are identified as melanin and hemoglobin [36, 37].

Because of the interaction of light with skin chromophores, information about skin chromophores can be extracted from the reflected light coming from skin. A digital camera acquires the reflected light and forms a digital color image. The digital color image is formed by bringing together three different spectral bands: red, green, and blue.

5.3.2.2 Melanin and Hemoglobin: Skin Color Model

Tsumura et al. reported that the spatial distribution of melanin and hemoglobin can be separated [36, 37]. The separation process is based on three assumptions. First, the process assumes that the optical density domain of the RGB channels is linear. Second, the spatial variations of skin color are assumed to be mainly due to two skin chromophores, melanin and hemoglobin. The third and final assumption states that the melanin and hemoglobin are independent, as depicted in Figure 5.8. Figure 5.9 illustrates the skin model based on these assumptions.

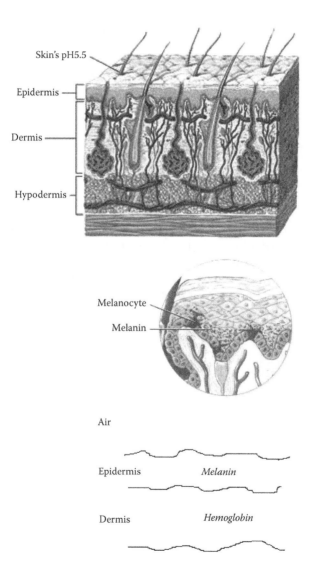

FIGURE 5.8
Skin chromophores.

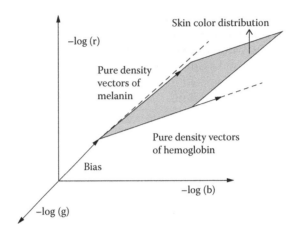

FIGURE 5.9
Skin color model (reproduced from Tsumura [36,37]).

5.3.3 VT-Scan Algorithm

5.3.3.1 Principal Component Analysis

Principal component analysis is an orthogonal linear transformation method. PCA transforms a dataset such that the greatest variance by any projection of the data come to lie on the first principal component of the new dataset, the second greatest variance on the second principal component, and so on [39]. In pattern recognition theory, a dataset with large variance is said to have large discriminatory power, thus it can be separated [15, 16].

A matrix A is a system of numbers with k rows and l columns.

$$A = \begin{pmatrix} a_{11} & a_{12} & \cdots & \cdots & a_{1l} \\ \vdots & a_{22} & & & \vdots \\ \vdots & \vdots & \ddots & & \vdots \\ \vdots & \vdots & & \ddots & \vdots \\ a_{k1} & a_{k2} & & & a_{kl} \end{pmatrix} \qquad (5.1)$$

Consider a $(k \times k)$ matrix, A. There are up to k eigenvalues in matrix A. For each eigenvalue there exists a corresponding eigenvector [40]. If there is a scalar λ and a vector γ such that

$$Ay = \lambda y, \qquad (5.2)$$

then component λ is an eigenvalue and γ is an eigenvector. It is proven that an eigenvalue, λ, is a root of the kth-order polynomial $|A - \lambda I| = 0$, where I is an identity matrix.

In applying PCA to the RGB images, the mean values of the color channels are initially subtracted to obtain zero mean variable data as follows:

$$R = R_0 - \mu_{R_0}, \ G = G_0 - \mu_{G_0}, \ B = B_0 - \mu_{B_0} \tag{5.3}$$

where R_0, G_0, and B_0 denote the image spectral band before subtraction and μ_{R_0}, μ_{G_0}, and μ_{B_0} denote the mean values of the respective image spectral bands.

Next, the covariance matrix is computed. The covariance matrix is a matrix of covariances between elements of datasets [41]. Covariance measures the degree of the linear relationship between variables. Larger values indicate higher redundancies, while smaller values mean low redundancies between the variables. The principal components can be derived from the covariance matrix and the correlation matrix. The general advantage of the covariance matrix is that the statistical inference is easier. This is true when all elements of the observations are measured in the same unit [42, 43].

$$\text{Cov} = \begin{bmatrix} C_{RR} & C_{GR} & C_{BR} \\ C_{RG} & C_{GG} & C_{BG} \\ C_{RB} & C_{GB} & C_{BB} \end{bmatrix} \tag{5.4}$$

where

$$C_{XX} = \frac{1}{N} \sum_{i=1}^{N} (X_i - \mu_i)^2 \tag{5.5}$$

$$C_{XY} = C_{YX} = \frac{1}{N} \left[\sum_{i=1}^{N} X_i Y_i \right] - \mu_x \mu_Y \tag{5.6}$$

$$\mu_X = \frac{1}{N} \sum_{i=1}^{N} X_i \tag{5.7}$$

$X, Y \in \{R, G, B\}$, N denotes the number of pixels in the image, and μ denotes the mean value.

The eigenvectors can be extracted from the covariance matrix by solving equation (5.8):

$$\text{Cov} = \gamma \lambda \gamma^T \tag{5.8}$$

where λ is a diagonal matrix representing eigenvalues of the covariance matrix, *Cov*, and γ is a matrix of eigenvectors of covariance matrix, *Cov*, arranged as columns.

The derived eigenvectors are used to transform of original (R, G, B) values. The vectors in the new space $[X_1 \ X_2 \ X_3]^T$ are obtained by

$$
\begin{bmatrix} X_1 \\ X_2 \\ X_3 \end{bmatrix} = \begin{bmatrix} \gamma_{11} & \gamma_{12} & \gamma_{13} \\ \gamma_{21} & \gamma_{22} & \gamma_{23} \\ \gamma_{31} & \gamma_{32} & \gamma_{33} \end{bmatrix} \begin{bmatrix} R \\ G \\ B \end{bmatrix}
\tag{5.9}
$$

where

$$
\begin{bmatrix} \gamma_{11} & \gamma_{12} & \gamma_{13} \\ \gamma_{21} & \gamma_{22} & \gamma_{23} \\ \gamma_{31} & \gamma_{32} & \gamma_{33} \end{bmatrix}
$$

are the eigenvectors of the covariance matrix.

It has been experimentally determined that after the application of PCA, the axis with the largest variance of PCA contained approximately 91% of the total variance [16]. Figure 5.10 shows an example of performing PCA on an RGB image.

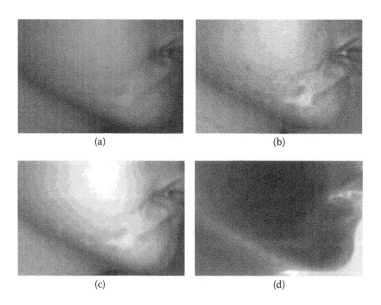

(a) (b) (c) (d)

FIGURE 5.10
Performing PCA on an RGB image: (a) original RGB image; (b) first principal component image; (c) second principal component image; (d) third principal component image.

It can be seen that the first principal component is an enhanced greyscale image showing more details of the original image as compared with other principal components.

5.3.3.2 Independent Component Analysis

Independent component analysis is a multivariate data analysis for source separation [44]. The basic model of ICA is a discrete time model in which sources are instantaneously mixed and the resulting mixture, possibly corrupted by noise, is observed. Expressing the source signals in vector form, $s(t) = [s_1(t), s_2(t), s_3(t), ..., s_M(t)]^T$, the N-dimensional observations, $x(t) = [x_1(t), x_2(t), x_3(t), ..., x_N(t)]^T$, are generated by a mixture corrupted by additive observation or sensor noise, $n(t)$, as follows:

$$x(t) = f(s(t)) + n(t) \tag{5.10}$$

where f is an unknown function.

The blind source separation technique aims to invert the function f and recover the sources. The mixing function, the noise, and the sources are unknown and therefore must be estimated. In ICA, it is assumed that the sources are mixed linearly by a mixing matrix, A. Thus observations are generated by

$$x(t) = A(s(t)) + n(t) \tag{5.11}$$

For simplicity, it is usually assumed that those vectors, s and n, have zero means. Generally ICA models assume the sources to be independent and the ICA models are noiseless [45, 46]. This can be written as

$$x = As \tag{5.12}$$

The ICA method is essentially used to recover the original sources from the observations by finding a separating matrix, W, to recover estimated sources, u, as follows:

$$u = Wx \tag{5.13}$$

The schematic of ICA is illustrated in Figure 5.11.

A fast and simple estimation of ICA called FastICA was developed by Oja et al. [47–49]. In this method, the independent components are obtained from maximizing non-Gaussianity. The basic intuitive lies on the central limit theorem [44]. The central limit theorem states:

Let $x_1, x_2, x_3, ...$ *be a sequence of random variables that are defined on the same probability space, share the same probability distribution, and are independent. The*

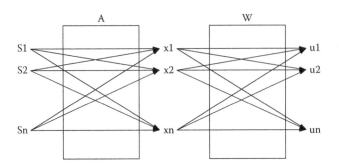

FIGURE 5.11
Independent component analysis.

convolution of all density functions of the random variables tends to be the nor-
mal density function (Gaussian distribution) as the number of density functions
increases, under the conditions stated above.

Assuming that the whitened data, $\tilde{x}$, is distributed according to the ICA
data model, the model can be formulated as

$$\tilde{x} = \tilde{A}s \tag{5.14}$$

The independent component, s, can be estimated by discovering the correct
linear combinations of the mixture variables. Let us denote a linear combination of the $\tilde{x}_i$,

$$a = w^T \tilde{x} = \sum_i w_i \tilde{x}_i \tag{5.15}$$

where w is a vector to be determined. By defining $q = \tilde{A}^T w$, this can be rewritten as

$$a = w^T \tilde{x} = w^T \tilde{A}s = q^T s \tag{5.16}$$

Thus a is a linear combination of s_i with weights given by q_i. The central
limit theorem ensures that the sum of two independent variables is more
Gaussian than the original variable. In other words $q^T s$ will be more Gaussian
than any of the s_i and it will become least Gaussian when it is one of the independent variables, s_i. Taking w as a vector that maximizes the non-Gaussianity
of $w^T \tilde{x}$, this vector will correspond to a q that has only one nonzero component. As a consequence, $w^T \tilde{x} = q^T s$ equals one of the independent components as shown in Figure 5.12. The figure shows that the mixed signal has a
density closer to the Gaussian compared with the independent component.

It is known that the Gaussian variable has the largest entropy among all
variables of equal variance [49]. In estimating the independent components, it

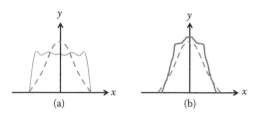

FIGURE 5.12
(a) Estimated density of one uniform independent component. (b) Marginal density of the mixed signal. Gaussian density (dashed curve) is given for comparison.

is necessary to have a quantification tool of non-Gaussianity of a random variable. To simplify the measurement, negentropy is used based on the information theoretic quantity of entropy. Negentropy, J, can be formulated as

$$J(x) = H(x_g) - H(x) \qquad (5.17)$$

where x_g is a Gaussian random variable of the same covariance matrix as x. Negentropy is always nonnegative, and it is zero if and only if x has a Gaussian distribution.

Negentropy is a well-justified statistical theory, but it is difficult to compute. General nonquadratic functions have been proposed to approximate negentropy [50, 51]. In general, the approximation of negentropy, $J(x)$, can be written as

$$J(x) \approx k_1 (E\{G^1(x)\})^2 + k_2 (E\{G^2(x)\} - E\{G^2(v)\})^2 \qquad (5.18)$$

where G^1 is odd and G^2 is even nonquadratic functions, k_1 and k_2 are positive constants, and v is a Gaussian variable of zero mean and unit variance. The variable x is assumed to have zero mean and unit variance. If we use only one nonquadratic function, G, the above equation becomes

$$J(x) \propto [E\{G(x)\} - E\{G(v)\}]^2 \qquad (5.19)$$

for practically any nonquadratic function G.

The performance of the approximation of negentropy depends on the nonquadratic function G. A good approximation of negentropy is obtained with an appropriate G. The following equations can be used as G:

$$G_1(u) = \tanh(a_1 u)$$

$$G_2(u) = u \exp(-u^2/2) \qquad (5.20)$$

$$G_3(u) = u^3$$

where $1 \le a_1 \le 2$ is some suitable constant.

FastICA is based on the fixed-point iteration method. The fixed-point iteration method is derived from the Newton iteration method. The maxima of $w^T \tilde{x}$ are obtained at certain optima of $E\{G(w^T \tilde{x})\}$. Following the Kuhn–Tucker conditions, the optima of $E\{G(w^T \tilde{x})\}$ under the constraint $\|w\|^2 = 1$ can be obtained at points where

$$E\{\tilde{x}G(w^T \tilde{x})\} + \beta w = 0 \qquad (5.21)$$

where β is some constant. To solve equation (5.21), the approximation of Newton's iteration becomes

$$w \leftarrow w - \frac{[E\{\tilde{x}G(w^T \tilde{x})\} + \beta \tilde{x}]}{[E\{G'(w^T \tilde{x})\} + \beta]} \qquad (5.22)$$

The above equation can be simplified as follows:

$$w \leftarrow E\{\tilde{x}G(w^T \tilde{x}) - E\{G'(w^T \tilde{x})\}w\} \qquad (5.23)$$

This is the basic fixed-point iteration in FastICA. The basic steps of FastICA are as follows:

1. Preprocessing the input data, x, into the whitened data $\tilde{x}$.
2. Initializing the weight vector w of the unit norm.
3. Let $w \leftarrow E\{\tilde{x}G(w^T \tilde{x}) - E\{G'(w^T \tilde{x})\}w\}$.
4. Let $w \leftarrow w / \|w\|$.
5. Go back to step 3 until convergence.

Convergence denotes that the previous and current values of w point in the same direction where their dot product is almost equal to 1. It is therefore not necessary for the vector to converge to a single point, so long as w and $-w$ point in the same direction. Figure 5.13 shows the outputs of PCA and ICA.

5.3.3.3 Segmentation

Region growing can be used to segment vitiligo areas [19]. Region growing is a segmentation method that groups pixels into regions or subregions based on predefined criteria [52, 53]. The method starts with a set of seed points and from these grow regions by appending to each seed those neighboring pixels that have properties similar to the seed (e.g., specific intensity levels).

Two samples of melanin images, representing intensity values of a vitiligo lesion and normal skin, are first obtained. Euclidian distance from every possible intensity level to the intensity of the sampled vitiligo lesion and normal skin are then calculated. Finally, the intensity value, I_{max}, is chosen as the intensity value that produces the optimum separation between vitiligo

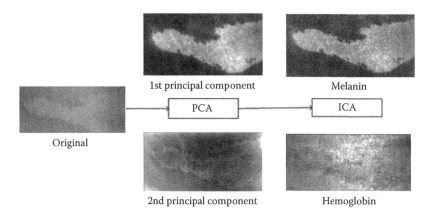

FIGURE 5.13
Outputs of PCA and ICA.

areas and normal skin areas. The method then takes the seed points from the position of the selected lesion sample. The defined lesion area is expanded from seed points by adding to each seed all neighboring pixels that have intensity lower than I_{max}.

5.3.3.4 Repigmentation Measurement

Repigmented skin is skin that has a colors similar to normal skin. However, the areas may be dispersed and too small to be easily discerned visually. The above segmentation process allows the differences in the vitiligo areas between skin images before and after treatment due to repigmentation to be objectively determined. The difference in percentage represents the repigmentation progression of a particular body region.

The calculation is explained as follows. Let $a(K, L)$ be the logical image where the vitiligo lesion and normal skin areas are represented by 1 and 0, respectively. $a(K, L)$ is defined as a processed image after vitiligo segmentation. The vitiligo area, $A_{vitiligo}$, is measured as follows:

$$A_{vitiligo} = \sum_{i=0}^{K} \sum_{j=0}^{L} a(i, j) \tag{5.24}$$

5.3.3.5 Overall Flowchart

The process flow of the vitiligo analysis system (VT-Scan) comprises preprocessing (PCA and whitening), ICA estimation, vitiligo segmentation, and repigmentation measurement.

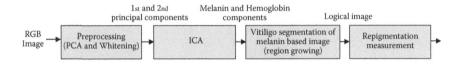

FIGURE 5.14
VT-Scan for repigmentation measurement of vitiligo lesion images [10].

5.4 Results and Analysis

5.4.1 Validation

In an observational study, forty-one digital images of vitiligo lesions from eighteen patients were obtained. The images obtained were from different body areas (head, nine; upper limbs, eight; trunk, fifteen; lower limbs, nine). To reduce artifacts due to specular reflections, the images underwent low-pass Gaussian filtering [54].

The validation is performed by comparing a manually segmented vitiligo lesion area of an image with the segmented vitiligo lesion area obtained using our method. The manual segmentation of vitiligo lesions is performed to ensure the objectivity of the result. Figure 5.15 illustrates the original vitiligo lesion, the manual segmentation output, and the computerized segmentation output.

The accuracy, sensitivity, and specificity parameters for each segmented image are then measured to evaluate the method and also to validate its use in clinical settings [55, 56]. The sensitivity parameter measures the proportion of actual vitiligo lesion area that is correctly identified. The specificity parameter measures the proportion of normal skin that is correctly identified. The accuracy parameter gives the overall performance.

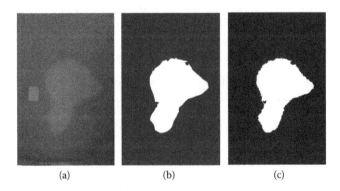

FIGURE 5.15
(a) Skin image. (b) Manual segmentation (ground truth). (c) Segmentation by VT-Scan.

Sensitivity is determined by equation (5.25):

$$\text{Sensitivity} = \frac{TP}{TP + FN} \tag{5.25}$$

True positive (TP) is when the proposed method correctly classifies pixels as vitiligo lesion. False negative (FN) is when the proposed method incorrectly classifies pixels as normal skin.

Specificity is determined by equation (5.26):

$$\text{Specificity} = \frac{TN}{TN + FP} \tag{5.26}$$

True negative (TN) is when the proposed method correctly classifies pixels as normal skin. False positive (FP) is when the proposed system incorrectly classifies pixels as vitiligo lesion.

Accuracy is determined by equation (5.27):

$$\text{Accuracy} = \frac{TP + TN}{TP + TN + FP + FN} \tag{5.27}$$

Using the above formulas, the accuracy, sensitivity, and specificity parameters of vitiligo segmentation of the dataset were found as shown in Table 5.3.

The sensitivity, specificity, and accuracy values of the method are high. The average accuracy is high, with a mean of 0.9901 (standard deviation [SD] 0.0092 (95% CI 0.0065, 0.012). The average specificity is also high with a mean of 0.9973 (SD 0.0031) (95% CI 0.0022, 0.004). With its high level of sensitivity (0.9105 [SD 0.0161]), high specificity, and accuracy the method can identify and measure lesion and normal skin areas correctly. Consistent high levels of accuracy, sensitivity, and specificity from different body areas indicate that the method does not depend on body parts and hence can be used to assess vitiligo lesions on all areas of the body.

TABLE 5.3

Performance of VT-Scan

Body Regions	Number of Images	Accuracy	Sensitivity	Specificity
Head/neck	9	0.9902	0.9189	0.9975
Upper limb	9	0.9914	0.9055	0.9981
Trunk	15	0.9877	0.9302	0.9962
Lower limb	8	0.9929	0.8739	0.9982
Average	41	0.9901	0.9105	0.9973

5.4.2 Observational Study

5.4.2.1 Study Protocol of the Observational Study

An observational study was conducted in Hospital Kuala Lumpur. Fifteen (15) pediatric patients were enrolled in the study. Enrolled patients were asked to apply treatment twice daily for twenty-four weeks or until two weeks after the affected areas defined for treatment at baseline were completely repigmented, whichever came first. A physical sunscreen was prescribed for daytime use on vitiligo lesions on sun-exposed areas. Scheduled treatment visits were conducted once every six weeks (i.e., baseline, week 6, week 12, week 18, week 24/end of treatment). Follow-up visits were conducted at week 30 and week 36/end of follow-up to assess the stability of the repigmented areas, to detect further improvement in repigmentation, and to detect relapse of depigmentation after discontinuation of treatment. Additional visits were conducted as necessary, particularly if there were unexpected treatment side effects. Digital images were taken at each visit for repigmentation analysis.

All efficacy assessments were performed at baseline/day 1 and at all subsequent study visits. The primary efficacy endpoint was the extent of repigmentation in the response to treatment expressed as a mean percentage of repigmentation (MPR) as assessed by the computerized technique (MPR-digital) or by the PGA (MPR-PGA).

The VT-Scan software was used to analyze and measure the surface area of repigmentation at baseline and week 24 (end of treatment). The digital images were taken under standardized lighting and position conditions. A representative vitiligo lesion from each body region selected as a treatment area was imaged for subsequent analysis. The difference in the surface area between baseline/day 1 and week 24/end of treatment as calculated by the VT-Scan analysis technique was expressed as a percentage of repigmentation in each vitiligo lesion. This percentage represents the repigmentation response for a particular body region.

Example:

Vitiligo surface area at baseline from body region A = 50 cm^2
Vitiligo surface area at week 24/end of treatment = 25 cm^2
Repigmented skin surface area = 50 cm^2 – 25 cm^2 = 25 cm^2
Percentage of repigmentation = 25/50 = 50% for A

For patients who had vitiligo affecting several body regions, the average percentage of repigmentation response for all the representative lesions was calculated and this was the mean percentage of repigmentation (MPR-digital).

The MPR-PGA was recorded every six weeks from baseline/day 1 to week 24/end of treatment for each anatomical site or body region. This evaluation rated the change of the representative vitiligo lesions in each body region using a 5-point ordinal scale as shown in Table 5.4. The PGA was also recorded for the entire body at visit 5/end of treatment.

TABLE 5.4

The 5-Point Ordinal Scale of Vitiligo

Score	Repigmentation (%)
None	0
Mild	1–25
Moderate	26–50
Good	51–75
Excellent to complete	76–100

5.4.2.2 Findings of the Observational Study

Using the computerized method, a total of nine (60%) patients were found to have at least some repigmentation, three (20%) had no response at all and three (20%) had worsening of the disease with treatment. The mean repigmentation (MPR) in these children was 23.6% (SD 38.6%) (95% CI 2.2%, 45%) with a median of 16%. The repigmentation response ranged from −26% to 95%.

A total of four (27%) patients achieved mild overall repigmentation, two (13%) had moderate, and three (20%) achieved excellent overall repigmentation with an MPR greater than 75% (Figure 5.17).

Two of three patients who had increased depigmentation with treatment had active disease with an increase in the size of vitiligo patches during the study period. One patient had a lesion in the areolar area of the left breast

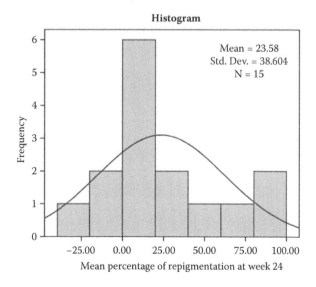

Histogram

Mean = 23.58
Std. Dev. = 38.604
N = 15

Mean percentage of repigmentation at week 24

FIGURE 5.16
MPR-digital [12].

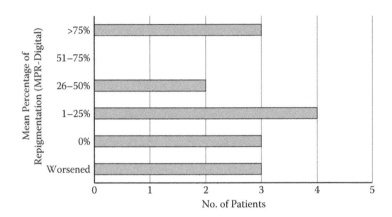

FIGURE 5.17
MPR-digital distribution.

that produced an uneven, nonplanar contour. This resulted in inconsistency in the surface area measurement by the computerized method.

With the PGA, ten of fifteen pediatric patients (67%) were observed to show some repigmentation with tacrolimus ointment 0.03%: four (27%) patients had mild response, four (27%) had moderate response, and two (14%) achieved at least 50% repigmentation (Figure 5.18).

Figure 5.19 shows the comparison between the two assessment methods. Both agreed that the majority of patients responded to treatment. However, there was discordance in all categories of repigmentation except for the mild response. According to the physician, none of the children's vitiligo was observed to have worsened with treatment, in contrast to the computerized method.

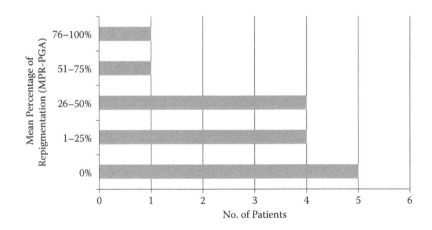

FIGURE 5.18
MPR-PGA distribution.

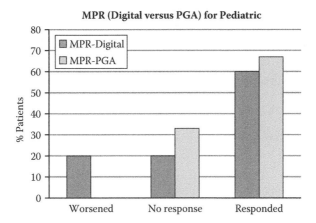

FIGURE 5.19
MPR-digital versus MPR-PGA.

The kappa test of consistency for the pediatric patients was 0.64 ($P < 0.0001$, 95% CI 0.334, 0.938), which shows good agreement between the two observations, and this is statistically significant.

5.5 Conclusion

Vitiligo is a pigmentary skin disorder resulting from abnormal melanin production due to the destruction of epidermal melanocytes. Vitiligo is a therapeutically challenging disease. Although a vast array of therapies are available, none produces optimal results, let alone a complete cure. To monitor the effectiveness of treatments, dermatologists need to examine the lesion directly or indirectly using digital photos. They analyze the disease by comparing patient's images before and after treatment. This requires a high degree of skill and experience, as the dermatologist has to be trained to provide an accurate assessment. Consequently it is possible to have different assessments due to varying degrees of experience.

Moreover, the progression of vitiligo treatment is very slow. The therapeutic response can be so subtle that the dermatologist may not be able to detect it. As a result, the disease is observed over a long period, typically every six months. Also, patients respond differently to the various vitiligo treatments and the therapeutic response of a particular treatment can be very different for different patients.

Skin color is due to the combination of skin chromophores, that is, melanin and hemoglobin. However, in digital imaging, color is created by combining three different spectral bands: red, green, and blue. To obtain accurate and

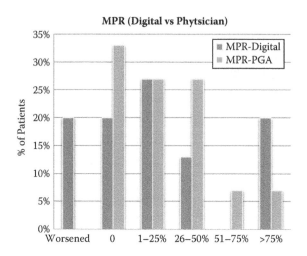

FIGURE 5.20
MPR-digital versus PGA [12].

objective measurements, the method must be able to determine skin areas due to skin chromophores (melanin and hemoglobin). VT-Scan is an image analysis software based on the skin color model developed by Tsumura et al. [36]. The model enables VT-Scan to determine accurately the chromophore composition of the skin from original RGB images, especially melanin. This enables dermatologists to accurately determine vitiligo and repigmentation areas.

We developed a method to analyze vitiligo lesions from digital color images by accurately segmenting vitiligo lesion areas using PCA, ICA, and segmentation. A validation study involving forty-one digital images of vitiligo lesions from eighteen patients was performed by comparing the manually segmented vitiligo lesion area of an image with the segmented vitiligo lesion area obtained using the system. The sensitivity, specificity, and accuracy were high. The average accuracy was extremely high with a mean of 0.9901 (SD 0.0092) (95% CI 0.0065, 0.012). The average specificity was also very high, with a mean of 0.9973 (SD 0.0031) (95% CI 0.0022, 0.004). The high level of sensitivity (0.9105 [SD 0.0161]), high specificity, and accuracy imply that the method can identify and measure lesion and normal skin areas correctly. This allows vitiligo monitoring to be performed within a shorter duration of six weeks compared with the typical three–six months.

An observational study involving fifteen pediatric patients was then conducted to evaluate the system. In the study we compared two assessment methods: an objective computerized digital imaging analysis technique, which was developed and validated concurrently with the clinical study, and a subjective method (the PGA). The outcomes from the two assessment methods used were compared to see whether there was any agreement. The kappa test of consistency between both assessments was 0.64 (P < 0.0001, 95% CI 0.334, 0.938), which shows substantial agreement between the two observations, and this is statistically significant. In summary, the computerized method can be effectively used as an objective tool for vitiligo monitoring.

5.A Appendix: VT-Scan MATLAB code

```
% VT-Scan ALGORITHM

function varargout = VTscan(varargin)
% Begin initialization code - DO NOT EDIT
gui_Singleton = 1;
gui_State = struct('gui_Name', mfilename,...

                'gui_Singleton', gui_Singleton,...
                'gui_OpeningFcn', @coba_OpeningFcn,...
                'gui_OutputFcn', @coba_OutputFcn,...
                'gui_LayoutFcn', [],...
                'gui_Callback', []);
if nargin && ischar(varargin{1})
    gui_State.gui_Callback = str2func(varargin{1});
end

if nargout
    [varargout{1:nargout}] = gui_mainfcn(gui_State, varargin{:});
else
    gui_mainfcn(gui_State, varargin{:});
end
% End initialization code - DO NOT EDIT

%- - Executes just before coba is made visible.
function coba_OpeningFcn(hObject, eventdata, handles, varargin)
% This function has no output args, see OutputFcn.
% hObject handle to figure
% eventdata reserved - to be defined in a future version of MATLAB
% handles structure with handles and user data (see GUIDATA)
% varargin command line arguments to coba (see VARARGIN)
% Choose default command line output for coba
handles.output = hObject;

% Update handles structure
guidata(hObject, handles);

% UIWAIT makes coba wait for user response (see UIRESUME)
% uiwait(handles.figure1);

%- - Outputs from this function are returned to the command line.
function varargout = VTscan_OutputFcn(hObject, eventdata, handles)
% varargout cell array for returning output args (see VARARGOUT);
% hObject handle to figure
% eventdata reserved - to be defined in a future version of MATLAB
% handles structure with handles and user data (see GUIDATA)

% Get default command line output from handles structure
varargout{1} = handles.output;

%- - Executes on button press in loadImage.
function loadImage_Callback(hObject, eventdata, handles)
```

```
% hObject handle to loadImage (see GCBO)
% eventdata reserved - to be defined in a future version of MATLAB
% handles structure with handles and user data (see GUIDATA)

[filename, pathname] = uigetfile({...
      '*.jpg;*.tif;*.gif;*.bmp;*.png', 'All image files(*.jpg,*.tif,*.gif,*.bmp,*.png)';
      '*.jpg;*.jpeg', 'JPEG files(*.jpg)';
      '*.gif', 'GIF files(*.gif)';
      '*.tif;*.tiff', 'TIFF files(*.tif)';
      '*.bmp', 'BMP files(*.bmp)';
      '*.png', 'PNG files(*.png)';
      '*.*', 'All Files (*.*)'}, 'Open an image');

im_original = imread([pathname, filename]);
figure('NumberTitle','off','Name','Please choose the green reference image'),
imshow(im_original);
figure('NumberTitle','off','Name','Please choose the green reference image'),Y =
imcrop(im_original);
close('NumberTitle','off','Name','Please choose the green reference image')

Y_green = imadjust(Y(:,:,2));
Y_bw = im2bw(Y_green,graythresh(Y_green));
Y_bw = medfilt2(Y_bw,[5 5]);
figure 3),imshow(Y_bw);
Y_out = sum(sum(Y_bw));
set(handles.textRef,'String',Y_out)

handles.RefText = Y_out;

figure('NumberTitle','off','Name','Please choose the lesion '),imshow(im_original);
figure('NumberTitle','off','Name','Please choose the lesion '),X=imcrop(im_original);
close('NumberTitle','off','Name','Please choose the lesion ')

figure('NumberTitle','off','Name','Please choose the lesion '),imshow(X);
figure('NumberTitle','off','Name','Please choose the lesion '),X2 = cutimage(X);
close('NumberTitle','off','Name','Please choose the lesion ')

handles.current_data = X2;
guidata(hObject,handles)

%- - Executes on button press in doICA.
function doICA_Callback(hObject, eventdata, handles)
% hObject handle to doICA (see GCBO)
% eventdata reserved - to be defined in a future version of MATLAB
% handles structure with handles and user data (see GUIDATA)

X = handles.current_data;
%save XX
%X = medfilt2(X,[5 5]);
[ril,a,b] = resatu(X);
ril2 = zeromean(ril');
[ICA_1,ICA_2] = buffer3(ril2,a,b);
handles.data1 = ICA_1(:,:,1);
handles.data2 = ICA_1(:,:,2);
guidata(hObject,handles)

%- - Executes on button press in doSatu.
function doSatu_Callback(hObject, eventdata, handles)
% hObject handle to doSatu (see GCBO)
% eventdata reserved - to be defined in a future version of MATLAB
% handles structure with handles and user data (see GUIDATA)
```

```
X = handles.data1;
X2 = mediancut(X);
X2 = tapis(X2);
X3 = sum(sum(X2));

set(handles.textPCA,'String',X3)
handles.PCAtext = X3;

Y _ out = handles.RefText;
ICA _ out = X3/Y _ out*1.13;

set(handles.textICA,'String',ICA _ out)
handles.data3 = X;
guidata(hObject,handles)

%- - Executes on button press in doDua.
function doDua _ Callback(hObject, eventdata, handles)
% hObject handle to doDua (see GCBO)
% eventdata reserved - to be defined in a future version of MATLAB
% handles structure with handles and user data (see GUIDATA)

X = handles.data2;
X2 = mediancut(X);
X2 = tapis(X2);
X3 = sum(sum(X2));

set(handles.textPCA,'String',X3)
handles.PCAtext = X3;
Y _ out = handles.RefText;
ICA _ out = X3/Y _ out*1.13;
set(handles.textICA,'String',ICA _ out)

handles.data3 = X;
guidata(hObject,handles)

%- - Executes on button press in doAuto.
function doAuto _ Callback(hObject, eventdata, handles)
% hObject handle to doAuto (see GCBO)
% eventdata reserved - to be defined in a future version of MATLAB
% handles structure with handles and user data (see GUIDATA)

X = handles.data3;
X _ out = tolol(X);
X _ sum = sum(sum(X _ out));
Y _ out = handles.RefText;
ICA _ out = X _ sum/Y _ out*1.13;
set(handles.textICA,'String',ICA _ out)
set(handles.textPCA,'String',X _ sum)
handles.currentdata = X _ out;
guidata(hObject,handles)

%- - Executes on button press in doInvert.
function doInvert _ Callback(hObject, eventdata, handles)
% hObject handle to doInvert (see GCBO)
% eventdata reserved - to be defined in a future version of MATLAB
% handles structure with handles and user data (see GUIDATA)

X _ out = handles.currentdata;
X _ out = logical(X _ out);
X _ out = ~X _ out;
X _ sum = sum(sum(X _ out));
Y _ out = handles.RefText;
```

```
ICA _ out = X _ sum/Y _ out*1.13;
set(handles.textICA,'String',ICA _ out)
set(handles.textPCA,'String',X _ sum)
figure,imshow(X _ out);
guidata(hObject,handles)

% CUTIMAGE FUNCTION

function out = cutimage(image)
X = image;
BW = roipoly(X);
X = im2double(X);
out(:,:,1) = BW.*X(:,:,1);
out(:,:,2) = BW.*X(:,:,2);
out(:,:,3) = BW.*X(:,:,3);

% TOLOL FUNCTION

function out = tolol(image)
a = im2double(image);
[j k] = size(a);
bare = max(max(a));
bare2 = bare/2;
b = a(a<bare2);
c = a(a> bare2);
satu = mean(b);
dua = mean(c);
image _ wrong = sqrt((a-double(satu)).^2);
image _ right = sqrt((a-double(dua)).^2);

out = ones(j,k);
for i = 1:j
   for p = 1:k
      x = image _ wrong(i,p);
      y = image _ right(i,p);
   if (x>y)
      out(i,p) = 1;
    else
      out(i,p) = 0;
    end

   end
 end

%out = tapis(out);
figure 11),subplot(1,2,2),imshow(out,[])
figure 11),subplot(1,2,1),imshow(image,[])

% BUFFER3 FUNCTION
function [imageICA,imageica] = buffer3(ril,a,b)
ril = ril';
[E D] = pcamat(real(ril),1,2);
[new,M,dm] = whitenv(real(ril),E,D);
out = fastica(ril,'approach','symm','g','tanh','whiteSig',new,'whiteMat',M,'dewhi
teMat',dm);
%out = out';
imageica = redua(real(out),a,b);
imageICA(:,:,1) = uint8(u8(imageica(:,:,1)));
imageICA(:,:,2) = uint8(u8(imageica(:,:,2)));
```

5.B Appendix: Kappa Interrater Agreement Analysis

Kappa is a statistical measure used in the interrater agreement analysis for the categorical data [57]. It can determine the agreement between two or multiple raters and for two or multiple categories. To determine the interrater agreement between two raters with two or multiple categories, Cohen's kappa is used. Given two 2D maskings of reference A and segmentation result B with n number of pixels that is designated into q number of categories, the overall agreement probability p_a is the proportion of pixels that both 2D maskings designated into the same category. It is common to all categories and given by equation (5.B.1).

$$p_a = \sum_{k=1}^{q} n_{kk}/n \qquad (5.B.1)$$

where n_{kk} is the number of pixels that is designated into the same category. Let n_{Ak} and n_{Bk} be the number of pixels that is designated into each category by set A and B, respectively. Then the chance agreement probability $p_{e|k}$ that is specific to each category is calculated using equation (5.B.2):

$$p_{e|k} = \sum_{k=1}^{q} n_{Ak} n_{Bk}/n^2 \qquad (5.B.2)$$

The Cohen's kappa $\hat{\gamma}_k$ is then calculated using equation (5.B.3):

$$\hat{\gamma}_k = (p_a - p_{e|k})/(1 - p_{e|k}) \qquad (5.B.3)$$

The kappa statistics value is from 0 to 1 and interpreted as the agreement strength between both 2D maskings as shown in Table 5.B.1.

TABLE 5.B.1

Agreement Strength of Kappa Statistics

Kappa	Agreement strength
<0.00	Poor
0.00–0.19	Slight
0.21–0.39	Fair
0.40–0.59	Moderate
0.60–0.79	Substantial
0.80–1.00	Outstanding

A kappa value of 0.60–0.79 is considered as having substantial agreement between both 2D maskings and a kappa value of 0.80–1.00 is considered to have outstanding agreement beyond chance.

References

1. Roberts N, Lesage M. Vitiligo: causes and treatment. *Pharmaceutical Journal* 2003;270:440–442.
2. Pajonk F, Weissenberger C, Witucki G, Henke M. Vitiligo at the sites of irradiation in a patient with Hodgkin's disease. *Strahlentherapie und Onkologie Organ der Deutschen Rontgengesellschaft* 2002;178(3):159–162.
3. Ingordo V, Gentile C, Iannazzone SS, Cusano F, Naldi L. Vitiligo and autoimmunity: an epidemiological study in a representative sample of young Italian males. *Journal of the European Academy of Dermatology and Venereology* 2011;25:105–109.
4. Jimbow K. Vitiligo. Therapeutic advances. *Dermatologic Clinics* 1998;16:399–407.
5. Halder RM, Young CM. New and emerging therapies for vitiligo. *Dermatologic Clinics* 2000;18:79–89, ix.
6. Boissy RE, Nordlund JJ. Vitiligo: current medical and scientific understanding. *Giornale italiano di dermatologia e venereologia organo ufficiale Societa italiana di dermatologia e sifilografia* 2011;146:69–75.
7. Parsad D, Pandhi R, Dogra S, Kumar B. Clinical study of repigmentation patterns with different treatment modalities and their correlation with speed and stability of repigmentation in 352 vitiliginous patches. *Journal of the American Academy of Dermatology* 2004;50:63–67.
8. Yang Y-S, Cho H-R, Ryou J-H, Lee M-H. Clinical study of repigmentation patterns with either narrow-band ultraviolet B (NBUVB) or 308 nm excimer laser treatment in Korean vitiligo patients. *International Journal of Dermatology* 2010;49:317–323.
9. Kostović K, Pastar Z, Pasić A, Ceović R. Treatment of vitiligo with narrow-band UVB and topical gel containing catalase and superoxide dismutase. *Acta Dermatovenerologica Croatica* 2007;15:10-14.
10. Fadzil MHA, Norashikin S, Suraiya HH, Nugroho H. Independent component analysis for assessing therapeutic response in vitiligo skin disorder. *Journal of Medical Engineering Technology* 2009;33:101–109.
11. Parsad D, Gupta S. Standard guidelines of care for vitiligo surgery. *Indian Journal of Dermatology, Venereology and Leprology* 2008;74(Suppl):S37–S45.
12. Shamsudin N. Efficacy and safety of tacrolimus ointment in vitiligo using a newly developed digital imaging analysis technique for evaluation of repigmentation progression. Universiti Kebangsaan Malaysia, 2009.
13. Hamzavi I, Jain H, McLean D, Shapiro J, Zeng H, Lui H. Parametric modeling of narrowband UV-B phototherapy for vitiligo using a novel quantitative tool: the Vitiligo Area Scoring Index. *Archives of Dermatology* 2004;140:677–683.
14. Fischer SAS. Colour segmentation for the analysis of pigmented skin lesions. *Signal Processing* 1997;443: 688–692.

15. Umbaugh SE, Moss RH, Stoecker WV, Hance GA. Automatic colour segmentation algorithms with application to skin tumor feature identification. *IEEE Engineering in Medicine and Biology Magazine* 1993;12(3):75–82.
16. Umbaugh SE, Wei YS, Zuke M. Feature extraction in image analysis. A program for facilitating data reduction in medical image classification. *IEEE Engineering in Medicine and Biology Magazine* 1997;16(4):62–73.
17. Cho K, Jang J, Hong K. Adaptive skin-colour filter. *Pattern Recognition* 2001; 34: 1067–1073.
18. Tomaz F, Candeias T, Shahbazkia H. Improved automatic skin detection in colour images. *Proceedings VIIth Digital Image Computing: Techniques and Applications: Proceedings of the VIIth Biennial Australian Pattern Recognition Society Conference, DICTA 2003*. Collingwood, Victoria, Australia: CSIRO, 2003:10–12.
19. van Geel N, Vander Haeghen Y, Ongenae K, Naeyaert J-M. A new digital image analysis system useful for surface assessment of vitiligo lesions in transplantation studies. *European Journal of Dermatology* 2004;14:150–155.
20. De Dios JJ, Garcia N. Fast face segmentation in component colour space. *2004 International Conference on Image Processing*, vol. 1. Washington, DC: IEEE, 2004.
21. Jones MJ, Rehg JM. Statistical colour models with application to skin detection. *IEEE Computer Society Conference on Computer Vision and Pattern Recognition*, vol. 1. Washington, DC: IEEE, 1999:274–280.
22. Caetano TS, Barone DAC. A probabilistic model for the human skin colour. *Proceedings of the 11th International Conference on Image Analysis and Processing*. Washington, DC: IEEE, 2001:279–283.
23. Zhu X, Yang J, Waibel A. Segmenting hands of arbitrary colour. *Proceedings of the Fourth IEEE International Conference on Automatic Face and Gesture Recognition*. Washington, DC: IEEE, 2000:446–453.
24. Chang F, Ma Z, Tian W. A region-based skin colour detection algorithm. *Lecture Notes in Computer Science* 2007;4426:417–424.
25. Kim J, Kim H. Multiresolution-based watersheds for efficient image segmentation. *Pattern Recognition Letters* 2003;24:473–488.
26. Phung SONLAM, Bouzerdoum A. A new image feature for fast detection of people in images. *International Journal of Information and System Sciences* 2007;3:383–391.
27. Seow M, Asari VK. Recurrent network as a nonlinear line attractor for skin colour association. *Lecture Notes in Computer Science* 2004;3173:870–875.
28. Mostafa L, Abdelazeem S. Face detection based on skin colour using neural networks. ICGST International Conference on Graphics, Vision and Image Processing, Cairo, Egypt, December 2005, pp. 19–21.
29. Takiwaki H, Shirai S, Kanno Y, Watanabe Y, Arase S. Quantification of erythema and pigmentation using a videomicroscope and a computer. *British Journal of Dermatology* 1994;131:85–92.
30. Kelly KM, Choi B, McFarlane S, Motosue A, Jung B, Khan MH, Ramirez-San-Juan JC, Nelson JS. Description and analysis of treatments for port-wine stain birthmarks. *Archives of Facial Plastic Surgery* 2005;7:287–2931.
31. Jung B, Choi B, Durkin AJ, Kelly KM, Nelson JS. Characterization of port wine stain skin erythema and melanin content using cross-polarized diffuse reflectance imaging. *Lasers in Surgery and Medicine* 2004;34:174–181.
32. Cotton SDO. Developing a predictive model of human skin colouring. *Proceedings SPIE 2708, Medical Imaging* 1996;2708:814–825.

33. Cotton S. Noninvasive skin imaging system. *Lecture Notes in Computer Science* 1997;1230:501–506.

34. Preece S, Cotton S, Claridge E. Imaging the pigments of human skin with a technique which is invariant to changes in surface geometry and intensity of illuminating light. In: Barber D, ed. *Proceedings of Medical Image Understanding and Analysis*. Malvern: British Machine Vision Association, 2003:145–148.

35. Claridge E, Cotton S, Hall P, Moncrieff M. From colour to tissue histology: physics-based interpretation of images of pigmented skin lesions. *Medical Image Analysis* 2003;7:489–502.

36. Tsumura N, Haneishi H, Miyake Y. Independent-component analysis of skin colour image. *Imaging* 2007;59:831–860.

37. Tsumura N, Haneishi H, Miyake Y. Independent component analysis of spectral absorbance image in human skin. *Optical Review* 2000;7:479–482.

38. Anderson RR, Parrish JA. The optics of human skin. *Journal of Investigative Dermatology* 1981;77:13–19.

39. Anthony D, Hines E, Barham J, Taylor D. A comparison of image compression by neural networks and principal component analysis. *Proceedings of the International Joint Conference on Neural Networks*. Hove, UK: Psychology Press, 1990.

40. Hadley SW, Mark BL, Vannelli A. An efficient eigenvector approach for finding netlist partitions. *IEEE Transactions on Computer-Aided Design* 1992;11:885–892.

41. Borgognone MG, Bussi J, Hough G. Principal component analysis in sensory analysis: covariance or correlation matrix? *Food Quality and Preference* 2001;12:323–326.

42. Joliffe IT, Learmouth JA, Pierce GJ, Santos MB, Trendafilov N, Zuur AF, Ieno EN, Smith GM. Principal component analysis applied to harbour porpoise fatty acid data. In Gail M, Krickeberg K, Sarnet J, Tsiatis A, Wong W, eds., *Analysing ecological data*. New York: Springer, 2007:515–528.

43. Joliffe IT, Morgan BJ. Principal component analysis and exploratory factor analysis. *Statistical Methods in Medical Research* 1992;1:69–95.

44. Hyvärinen A, Karhunen J, Oja E. *Independent component analysis*. New York: Wiley-Interscience, 2001:481.

45. Comon P. Independent component analysis, a new concept? *Signal Processing* 1994;36:287–314.

46. Boscolo R, Pan H, Roychowdhury VP. Independent component analysis based on nonparametric density estimation. *IEEE Transactions on Neural Networks* 2004;15:55–65.

47. Oja E. Convergence of the symmetrical FastICA algorithm. *Proceedings of the 9th International Conference on Neural Information Processing*, vol. 3. Washington, DC: IEEE, 2002:1368–1372.

48. Oja E, Yuan Z. The FastICA algorithm revisited: convergence analysis. *IEEE Transactions on Neural Networks* 2006;17:1370–1381.

49. Hyvärinen A, Cristescu R, Oja E. A fast algorithm for estimating overcomplete ICA bases for image windows. *International Joint Conference on Neural Networks*. Washington, DC: IEEE, 1999:894–899.

50. Li H, Adali T. A class of complex ICA algorithms based on the kurtosis cost function. *IEEE Transactions on Neural Networks* 2008;19:408–420.

51. Novey M, Adali T. Complex ICA by negentropy maximization. *IEEE Transactions on Neural Networks* 2008;19:596–609.

52. Shih F, Cheng S. Automatic seeded region growing for colour image segmentation. *Image and Vision Computing* 2005;23:877–886.
53. Maglogiannis I. Automated segmentation and registration of dermatological images. *Journal of Mathematical Modelling and Algorithms* 2003;3:277–294.
54. Liang Z. Implementation of linear filters for iterative penalized maximum likelihood SPECT reconstruction. *IEEE Transactions on Nuclear Science* 1991;38:606–611.
55. Fadzil MHA, Izhar LI, Nugroho H, Nugroho HA. Analysis of retinal fundus images for grading of diabetic retinopathy severity. *Medical and Biological Engineering and Computing* 2011;49:693–700.
56. Fawcett T. Introduction to receiver operator curves. *Pattern Recognition Letters* 2006;27:861–874.
57. Haddad AL, Matos LF, Brunstein F, Ferreira LM, Silva A, and Costa D Jr. A clinical, prospective, randomized, double-blind trial comparing skin whitening complex with hydroquinone vs. placebo in the treatment of melasma. *International Journal of Dermatology* 2003;42:153–156.

6

Quantitative Assessment of Ulcer Wound Volume

Ahmad Majdi A. Rani, Ahmad Fadzil Mohamad Hani,
Nejood El-Tegani, Evan Chong, and Ankur Sagar

CONTENTS

6.1 Ulcers

6.1.1 Chronic Ulcers

Ulcer refers to a discontinuity of the skin exhibiting complete loss of the epidermis that is not short lived. The duration of the ulcer could last from a few weeks to even a few years. Patients suffering from such chronic skin ulcers face a huge loss of quality of life. The patients not only have to endure pain, but also time-consuming outpatient treatment and costs. Leg ulcers are very common, and based on the statistics, 3.5% of all adults in the United States suffer from venous leg ulcers [1]. Chronic ulcerative skin lesions affect about 1.5% of the European population and represent an important medical and social problem [2].

Ulceration is most commonly seen on the lower extremities. Ulcers generally occur in several areas, including the mouth, gastrointestinal tract, skin, and corneas. In dermatology, ulcer refers to any discontinuity of the skin that appears as inflamed tissue with reddened skin. This kind of ulcer can occur anywhere on the body. Four types of common ulcers are depicted in the Figure 6.1.

An increase in the number of patients with chronic wounds has been recorded with the population advancing in age and increasing in weight, and with the resultant comorbidities of diabetes and venous insufficiency. According to the estimates, approximately 1% of the world's population will develop a leg ulcer during their life [3]. In the United States alone, chronic wounds affect 3 to 6 million patients and treating these wounds costs an estimated of $5 to $10 billion each year.

Chronic ulcers are wounds that fail to heal within the estimated period. Wound healing is divided into three stages: inflammation, tissue formation, and tissue remodeling.

Inflammation is the first phase of wound healing. Under normal conditions, it usually lasts for four–six days. The main processes of inflammation

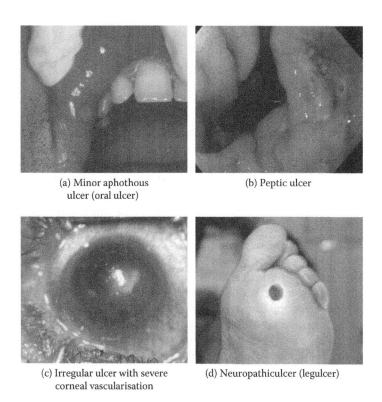

(a) Minor aphothous
ulcer (oral ulcer)

(b) Peptic ulcer

(c) Irregular ulcer with severe
corneal vascularisation

(d) Neuropathiculcer (legulcer)

FIGURE 6.1
Four common types of ulcer: (a) minor aphthous ulcer (oral ulcer); (b) peptic ulcer; (c) irregular ulcer with severe corneal vascularization; and (d) neuropathic ulcer (leg ulcer).

are vasoconstriction, hemostasis, and vascular dilation with increased capillary permeability, chemotactic growth factor, and phagocytosis. The second phase of wound healing begins about four–five days after the wounding and lasts for few weeks. It is the most important phase in the process of wound healing. The main processes in this phase are angiogenesis, granulation tissue formation, reepithelialization, and extracellular matrix formation. The tissue formation phase is also known as the proliferative phase. Eventually a continuous process of dynamic equilibrium between the synthesis of new stable collagen and the lysis of old collagen takes place. This process is called tissue remodeling and can take up to two years. Lower limb ulceration tends to be recurrent and can become a chronic wound if the wound does not heal in an orderly fashion and in the normal period. Since chronic ulcers are hard to heal, monitoring wound healing progress becomes crucial.

Leg ulcers commonly occur during late middle or old age due to the chronic venous insufficiency (CVI), chronic arterial insufficiency, or peripheral sensory neuropathy, or a combination of these factors. Leg ulcers result in long-term morbidity and often do not heal unless the underlying cause is corrected.

6.1.2 Leg Ulcers

In pathology, a wound refers to an injury that causes discontinuity of the anatomical structure and function of skin tissue. When the skin is injured due to a trauma, it goes through a series of overlapping processes to repair the damaged tissues. Chronic wounds or ulcers occur when the injured tissue does not follow a normal course of healing within the expected period of time due to untreated underlying etiologies or improper wound management. Nonhealing ulcers can remain for years, causing pain and discomfort to patients and exposing them to the risk of infection and limb amputation.

According to recent studies, more than 3% of the adult population has the potential to develop chronic wounds during their lifetime, with a significantly increased prevalence in the elderly [3]. The prevalence of chronic wounds may increase significantly due to the increase in the population's age or underlying etiologies such as diabetes and venous and arterial insufficiencies. Leg ulcers are common chronic wounds that are found on the lower extremities below the knee and mostly affect people age sixty years and older.

Leg ulcers are classified into four types: venous, arterial, combined venous and arterial, and neuropathy. Each ulcer type has its own symptoms, causes, shapes, and locations. Most leg ulcers are either vascular or diabetic and affect 1% of the adult population and 3.6% of people older than sixty-five years [5].

6.1.2.1 Venous Ulcers

Increasing patient age, obesity, leg injury (fractures), deep vein thrombosis (DVT), and phlebitis are associated with the increase in the prevalence of venous ulcers, which is approximately 1%. Among the symptoms that occur are limb heaviness, swelling associated with standing, which worsens in the evening, and pain. Usually venous ulcers are associated with at least one of the symptoms of chronic venous insufficiency (CVI) (see Figure 6.2). Venous ulcers may appear as a single ulcer or in multiples. Venous ulcers are usually found in the area supplied by an incompetent perforating vein, such as on the medial lower aspect of the calf, especially over the malleolus, and often can be as large as the circumference of the entire lower leg.

The ulcers are usually painful, well-defined, irregular-shaped, and relatively shallow with a sloping border with the base covered by fibrin and necrotic material, which are always due to secondary bacterial colonization. A morbidly obese individual has a risk of developing stasis ulcers.

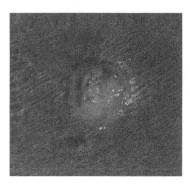
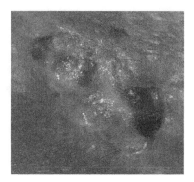

FIGURE 6.2
Venous insufficiency [4]. (a) Two coalescing ulcers with a necrotic base. (b) A giant ulcer with scalloped borders and a beefy red base.

Long-standing venous ulcers are associated with the risk of developing squamous cell carcinoma (SCC), shown in Figure 6.3.

6.1.2.2 Arterial Ulcers

Arterial ulcers have an age-adjusted prevalence of 12% and they are usually seen in patients with peripheral arterial disease. Patients commonly suffer from symptoms of intermittent claudication and pain, even at rest. As the disease progresses, they tend to have a characteristically painful leg at night, which is often severe. The pain may worsen when the legs are elevated. The common locations for arterial ulcers are over sites of pressure and trauma, such as the pretibial supramalleolar and toes. Arterial ulcers are painful and have a punched out appearance with sharply demarcated borders. An example of chronic arterial insufficiency is shown in Figure 6.4.

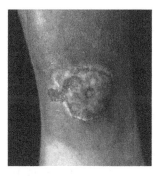

FIGURE 6.3
Squamous cell carcinoma in a chronic venous ulcer [4].

FIGURE 6.4
Chronic arterial insufficiencies with punched out edges and irregular borders [4].

Sometimes tendons can be seen under the base of the ulcers and these are covered by slough tissues. However, they have a minimal amount of exudation. Common findings in a patient with arterial ulcers are loss of hair on the feet and lower legs, shiny atrophic skin, and diminished or absent pulse. In contrast to venous ulcers, arterial ulcers do not have stasis pigmentation and lipodermatosclerosis.

Martorell's ulcer is a special type of arterial ulcer that is associated with labile hypertension and lacks signs of atherosclerosis obliterans. This is a very painful ulcer located on the anterior lateral lower leg, which usually starts with black eschar with surrounding erythematous tissue. It has a punched out appearance with sharply demarcated borders and surrounding erythematous tissue after sloughing of necrotic tissue.

6.1.2.3 Mixed Ulcers

These ulcers have a combination of signs and symptoms of both venous and arterial insufficiency and ulceration, as the patients usually have both CVI and atherosclerosis obliterans. Among the symptoms are intermittent claudication, pain when elevated or when the leg is put in a dependent position, both pallor and cyanosis of the foot, stasis dermatitis, and lipodermatosclerosis. The ulcers have both a sloped and punched out appearance that reaches down to the tendons, as seen in Figure 6.5.

6.1.2.4 Neuropathic Ulcers

These types of ulcers are usually found on the soles, toes, and heels and are commonly associated with diabetes. Figure 6.6 shows a neuropathy ulcer on the sole. Among the symptoms are paraesthesia, pain, and anesthesia of the leg and foot. Patients are usually unaware of their ulcers, as they have lost feeling in their feet.

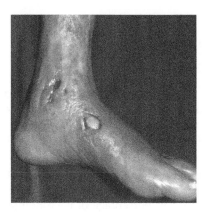

FIGURE 6.5
Chronic combined arterial and venous ulcers [4].

"Diabetic foot" is a common term associated with peripheral neuropathy. However, other conditions that can contribute to diabetic feet are angiopathy, atherosclerosis, and infection. Usually these occur together to cause diabetic foot. Diabetic neuropathy results from a combination of motor and sensory deficit that causes patients to be unable to feel any pain. Weakness and distal muscle wasting are due to motor neuropathy. Neurotropic ulcers over the bony prominences of feet, usually on the great toe and sole, are the result of sensory neuropathy. Osteomyelitis is one of the complications that may arise as the ulcers are surrounded by a ring of callus and which can extend the joint and bone.

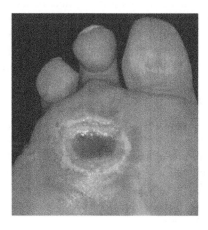

FIGURE 6.6
Diabetic neuropathic ulcer on the sole [4].

6.2 Assessment of Leg Ulcers

6.2.1 Current Clinical Practice

In clinical practice, the evaluation of leg ulcers is highly dependent on the skill of the dermatologist as the assessments are visual observations and estimations of the ulcer wound geometric shape. Assessments such as the Leg Ulcer Measurement Tools (LUMT) serve as a guide to monitor the ulcer and treatment efficacy [5]. It is time consuming for dermatologist to perform all the assessment parameters proposed in the LUMT. Hence only four main criteria of evaluation are used in clinical practice: volume measurement, area measurement, percentage of granulation tissue, and percentage of necrotic tissue. As a consequence of wound healing, there will be tissue growth at the wound bed resulting in a decrease in wound volume followed by a slower decrease in wound perimeter [6].

Assessments are very subjective and may differ from one dermatologist to another, leading to inter- and intrarater variability. Hence a quantitative and objective measurement is crucial for the assessment in order to accurately monitor the healing process and treatment efficacy. Knowing which ulcers fail to follow the normal healing path will enable the dermatologist to choose an alternative treatment.

Apart from visual assessments, measurement tools, including rulers, acetate sheets, swabs, and saline, also help dermatologists achieve better assessments. The aforementioned criteria for the assessment of leg ulcers require different measurement tools. An objective and quantitative assessment method is crucial to determine the amount of healing in response to various treatments, medications, and disease processes.

There are more than 200 possible dressing materials and treatments available and new products are constantly promoted to dermatologists. Different dressings and treatment remedies are applied based on ulcer severity. In addition, the dressing material and treatment may not be suitable to a particular person due to a unique body structure. Nonetheless, the earlier the changes in wound parameters are determined, the earlier doctors can make clinical decisions on suitable treatments needed for wound healing.

6.2.2 Volume Measurement

An indicator of wound healing is a change in wound volume followed by a slow decrease in the perimeter and area [7]. As the ulcer heals, red granulation tissue starts to grow from the base of the ulcer, gradually replacing the black necrosis and yellow slough, filling the wound cavity and reducing its volume. However, volume measurement is not an easy process and the most common practice is to multiply the area by the wound depth through an intrusive measurement process.

For depth measurement, a sterile swab is inserted into the deepest area of the ulcer to estimate the depth of ulcer. The height parallel to the external

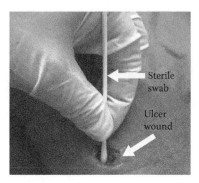

FIGURE 6.7
Technique for tracing the ulcer margin [8].

boundary of the ulcer is observed and marked in order to determine the ulcer's depth. The process of marking the height may be flawed due to parallax error on the part of the observer. In addition, the head of the swab may not be in the deepest part of the wound bed, which can lead to incorrect measurements. Figure 6.7 shows the technique for tracing the wound boundary using a sterile swab.

Another measuring tool, called a Kundin gauge, has been developed for depth measurement. The Kundin gauge also allows the measurement of length and width. Figure 6.8 shows the Kundin gauge that was developed for the measurement of three dimensions (length, width, and height).

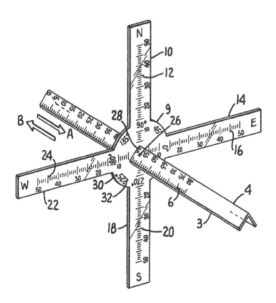

FIGURE 6.8
A Kundin gauge [9].

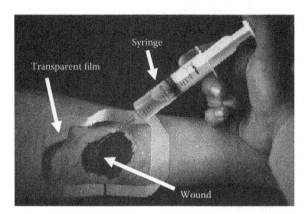

FIGURE 6.9
The saline injection technique.

However, this method is not very accurate and there are inherent errors. Wounds rarely appear as regular solids and do not have a uniform depth, width, and length [10]. Measuring the three dimensions of the wound and multiplying each of them could result in overestimating or underestimating the volume. These measurements are notoriously inaccurate due to irregular wounds and poor reproducibility of boundary selection.

Instead of measuring the volume by multiplying the three dimensions of wounds, there are two other methods used to determine the volume of the ulcer cavity: saline injection and mold making [11]. In the saline injection method, the ulcer is covered with a transparent film and then filled with saline using a syringe, as shown in Figure 6.9 [12, 13]. The amount of saline dispensed from the syringe shows the volume of the ulcer [8]. However, the wound may absorb some of the saline and the shape of the plastic cover may not be the same as the original healthy skin, affecting the accuracy of the method to determine ulcer volume.

Another technique of volume measurement is to fill the ulcer cavity with alginate or a silicon-based paste to make a mold. The material is then weighed to determine the amount of material used [13, 14]. The material is placed into a beaker, where the water displacement indicates the cavity volume.

In general, the above assessments of wound volume can be classified as intrusive. Some difficulties that can arise when intrusive methods are used include (1) the method can cause discomfort to patients and introduce a risk of wound infection; (2) overestimation or underestimation in volume can occur, as the filling process is based on visualization; and (3) these methods are not suitable for large wounds, wounds on limbs with large curvatures, and shallow wounds.

6.2.3 Wound Assessment Using the Digital Imaging Technique

Wound assessment using digital imaging can be performed using either two-dimensional (2D) or three-dimensional (3D) images. 2D-based assessment of wound color images typically characterizes tissue status using segmentation [15, 16]. This method can be used to measure the percentages of granulation tissue and necrotic tissue. Depth is not considered since there is no depth information in 2D images.

In the case of 3D-based wound measurements, prototypes based on the laser triangulation method [6, 17, 18], structured light technique [19, 20], and photogrammetry [21, 22] to obtain spatial measurements have been reported, but these have not been adopted in clinical practice.

6.2.3.1 Structured Light

The Measurement of Area and Volume Instrument System (MAVIS) is an instrument developed to measure the volume of skin wounds, pressure sores, and ulcers utilizing the principle of color-coded structured light [23]. Initially the operator has to define the boundary of the exposed skin surface manually by tracing the edge of the image. Tracing the edge is also used for measurement of the wound perimeter and area measurement. The second step is to reconstruct the healthy skin. The volume of wound is defined as the volume of the region sandwiched between the exposed skin surface and the original healthy skin surface, which is imitated using cubic spline interpolation. Cubic spline interpolation has a tendency to create a minimum curvature that is similar to the behavior of normal skin.

As shown in Figure 6.10, MAVIS is a huge system that is mounted on a trolley and weighs 110 kg. Although MAVIS produces better results compared with conventional measurement tools, it is not suitable for wounds that are located on high-curvature surfaces, such as lower limbs. Also, it is incapable of measuring wounds that are undermined or hollowed (see Table 6.1) due to the optical principle. Generally, volume measurements with a large area:volume ratio are less precise than those with smaller ratios.

6.2.3.2 Laser Triangulation

Derma is an integrated tool used to measure and monitor the evolution of skin lesions with a 3D system. It uses a laser triangulation 3D scanner to acquire wound geometric data and also to capture a red, green, and blue (RGB) image of the wound. Derma supports all the geometric measures, such as length (distance between critical points, perimeter, depth, and different subregions that are separated by color), surface area, and volume.

The selection of the border is based on a semiautomated manner. The border of the wound is first defined by the user or dermatologist. Next, the system improves the fit of a predefined border by considering the shape and color gradient of a selected area. The estimated healthy skin is then reconstructed to cover the wound (which is called a lesion tap). The lesion tap is

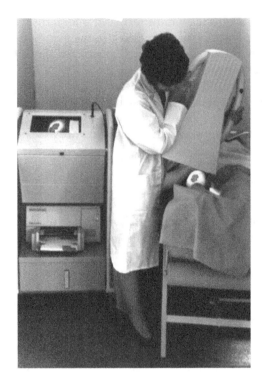

FIGURE 6.10
Trolley-mounted MAVIS equipment [1].

reconstructed by interpolating a surface over the proximity of wound border, as depicted in Figure 6.11.

In addition to volume measurement, Derma is also used to measure the percentage of granulation and necrotic tissue in the wound. Although Derma has shown promising results, it is not suitable for wounds in highly curved areas such as the lower limbs.

6.2.3.3 Photogrammetry

Photogrammetry technology is typically based on the process of recording, measuring, and interpreting photographic images [24]. Here, MAVIS is replaced by MAVIS II, which uses a reflex digital camera equipped with special dual-lens optics to record two half images from slightly different viewpoints [1]. The operating principle of MAVIS II is based on stereophotogrammetry, as shown in Figure 6.12. The distance is directly measured by the displacement between corresponding points in the stereo images. Although MAVIS II overcomes the weaknesses of MAVIS, such as size, cost, and time-consuming measurements, the method of healthy skin reconstruction is not suitable for wounds in highly curved areas.

TABLE 6.1

Common Wound Attributes, Descriptors, and their Schematic Diagrams

Attribute	Descriptor	Schematic
Boundary	Regular	
	Irregular	
Edge	Sloped	
	Punched Out	
	Undermined/hollowed/undercut	

(continued)

TABLE 6.1 (CONTINUED)

Common Wound Attributes, Descriptors, and their Schematic Diagrams

Attribute	Descriptor	Schematic
	Elevated	
Base	Homogeneous	
	Depressed	
Depth	Unit (mm)	x mm

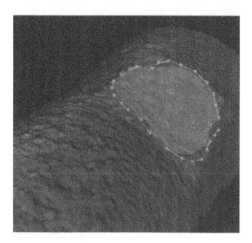

FIGURE 6.11
Wound covered with a thin layer of healthy skin for volume measurement [2].

The Medical Photogrammetric System (MEDPHOS) [21, 25, 26] was developed to provide 3D information, measurement, and reconstruction of wound surfaces, which are beneficial for wound assessment. The working principle of MEDPHOS is based on a combination of active and passive 3D data acquisition. A dot target projection generated from a slide projector serves as an active camera with a known calibration parameter. However, researchers have focused on development of the photogrammetry system, not the accuracy of volume computation.

ESCarre Analyse Lisibilité Evaluation (ESCALE) is another system designed for objective and accurate 2D and 3D measurements for the assessment of wound healing [27]. There are two major parts in ESCALE: (1) 3D model reconstruction and volume measurement based on photogrammetry, and (2) tissue classification [28, 29]. The process flow of the ESCALE system is shown in Figure 6.13.

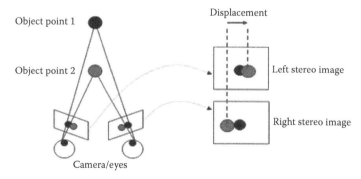

FIGURE 6.12
Operating principles of stereophotogrammetry [1].

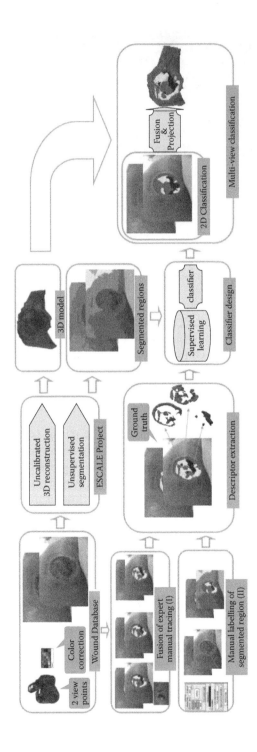

FIGURE 6.13
ESCALE wound assessment tool [29].

6.3 Development of Objective Volume Measurement of Wounds

6.3.1 CAD Modeling of Ulcer Attributes

Ulcers can appear in various regular or irregular shapes of different sizes. Regular shapes are round, ellipse, or even an almost rectangular. According to Hani et al. [30], ulcers can be described using several wound attributes, which are shown in Figure 6.14. These wound attributes can be classified into four classes: wound boundary, wound edges, wound base, and wound depth.

In addition, each of these attributes can be further described with various descriptors, as shown in Table 6.1, to further assist in wound assessment.

With a combination of these attributes and descriptors, seventeen ulcer wound computer-aided design (CAD) models with regular and irregular shapes were created. The volume of each ulcer wound cavity was measured and recorded from the software and will be used as benchmarks for validation of volumes obtained from prototype models.

6.3.2 Ulcer Wound Prototype Models

The prototypes of the seventeen ulcer wound models were created using a Multi-Jet modeling 3D printer. The drawing file format of the ulcer wound CAD models was converted into STL format for 3D printing of the ulcer wound prototypes. Using solid object printer software, the STL format file was imported and the CAD models were arranged for printing as shown in Figure 6.15. The Multi-Jet modeling printer uses the STL file to print layer by layer from the bottom moving upward until the ulcer wound prototype models were complete.

Figure 6.16 shows the prototypes produced by the solid object printer. There is one postprocessing function for the Multi-Jet modeling, which is removal of support material. The support material is built below the parts

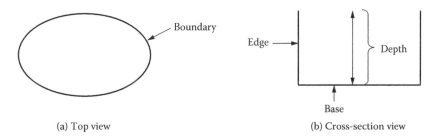

(a) Top view   (b) Cross-section view

FIGURE 6.14
Schematic diagram of ulcer attributes: (a) top view; (b) cross-section view.

FIGURE 6.15
Arrangement of the 3D model in a solid object printer.

and the gap between materials. It can be scrubbed off manually using a scrapper or sandpaper. The volume of each ulcer wound cavity from the prototype models was then measured and recorded.

6.3.3 Volume Computation of Wound Models

Three different approaches were selected to measure the cavity volume of the prototype model. The first method uses the conventional method for volume measurement either by using saline injection or filling with mold material. Both techniques will give similar reliable results. The second and

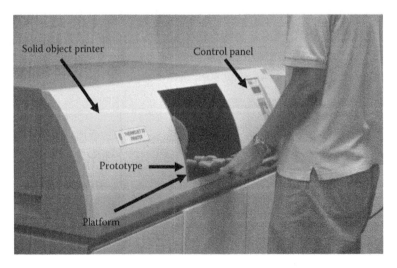

FIGURE 6.16
Prototype produced by a solid object printer.

third methods are using reverse engineering tools—laser triangulation and structured light—to obtain geometric information about each prototype. Laser triangulation and structured light data acquisition techniques are two of the most common techniques in reverse engineering an object. These two techniques were adopted in this work to scan and capture the 3D skin surface images of the ulcers. The two most common algorithms used for volume computations—midpoint projection and convex hull approximation—were used to reconstruct the estimated surface and compute the volumes of the ulcer wound cavities.

6.3.3.1 Surface Reconstruction Using Midpoint Projection

Before computing wound volume, a solid must be reconstructed out of a wound surface scan. Here, a solid is reconstructed by connecting all the triangular faces to a midpoint, creating many tetrahedra. The midpoint is calculated from a number of points selected at the edges of the wound. Figure 6.17 shows the midpoint calculated from a number of edges at the boundary.

The volume of the cavity or wound is computed by summing up the volume of all the tetrahedra. Projecting surfaces to a plane or reference point is useful for volume reconstruction. The shape of the model under construction is the key in selecting the appropriate projection method. In the case of wound models, the reconstruction of a projected wound surface to an edge point, reference plane, or a point interpolated at the top surface will lead to an approximation of wound volume. However, one of the issues in developing the algorithm is to obtain reproducible results of the wound volume. If a point at the boundary is used as a reference point for solid reconstruction, it is not possible to landmark the wound boundary and measure the wound using the same point in each assessment. Projecting the surface faces to one edge point, which is difficult to identify in each assessment (landmark), leads to results that are not always reproducible. Calculating a midpoint from several points at the wound boundary will produce more reproducible results.

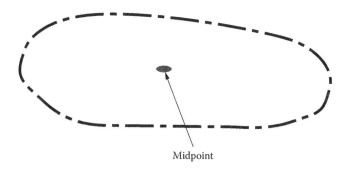

Midpoint

FIGURE 6.17
Midpoint calculation.

The midpoint is obtained from a set of points selected at the wound boundary. The SELECT3D MATLAB tool can be used to determine the selected point P in the 3D data space. P is a point on the first patch or surface face intersected along the selection ray. It returns the closest face vertex. Figure 6.18 shows the results of constructing solids from a model and a

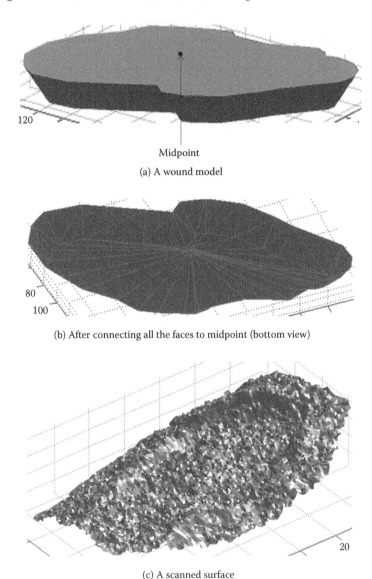

(a) A wound model

(b) After connecting all the faces to midpoint (bottom view)

(c) A scanned surface

FIGURE 6.18
Midpoint solid reconstruction for a model and a skin surface: (a) a wound model and (b) its reconstructed solid; (c) the wound surface scan and (d) its solid model. (*continued*)

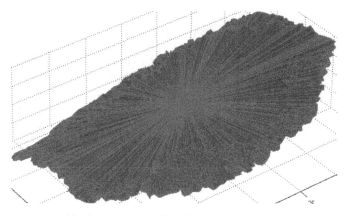

(d) After connecting all the faces to midpoint

FIGURE 6.18
(*continued*) Midpoint solid reconstruction for a model and a skin surface: (a) a wound model and (b) its reconstructed solid; (c) the wound surface scan and (d) its solid model.

real scan using midpoint solid reconstruction. Figure 6.18(a) shows a wound model and Figure 6.18(b) shows its reconstructed solid using midpoint projection. Figure 6.18(c) shows the wound surface scan, and Figure 6.18(d) shows its solid model created using midpoint projection.

In unusual wounds (e.g., an L-shaped wound) in which the midpoint might lie outside the surface, as shown in Figure 6.19, recalculation of the midpoint

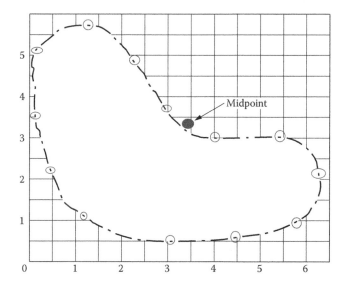

FIGURE 6.19
An L-shaped wound.

is required. The same z value can be used for the new midpoint and the location of the x and y coordinates should be altered.

Modification or recalculation of the midpoint allows the user to select a point inside the surface. The x and y coordinates are obtained from the user selection while the z value from midpoint calculation is still valid.

This algorithm is sensitive to the existence of holes at the surface because all the faces of the surface in the triangular mesh are used in volume computation. Even though surface registration eliminates gaps at the surface scan, some holes at the surface may be left unfilled. In order to use this algorithm for volume computation, all the holes at the surface must be filled. This can be performed as a preprocessing step.

When using this method for solid reconstruction and volume computation, points around the surface do not need to be dense. In the case of the surface scan, the number of vertices in the surface can be reduced using the decimation process, which will lead to a reduction in algorithm execution time. Decimation refers to the process of reducing the number of triangles while preserving the original surface shape.

The flow chart of reconstructing wound models using midpoint projection is shown in Figure 6.20.

The algorithm from the flow chart in Figure 6.20 is outlined as follows:

1. Input the triangulated surface.
2. Translate the surface to positive x, y, and z coordinates.
3. Select several points on the wound outer boundaries.
4. Average all the points to obtain midpoint x, y, and z values.
5. Connect all triangular faces to the midpoint creating a collection of tetrahedra (the midpoint will represent the fourth vertex in all the tetrahedra in the model).
6. Compute the volume of the wound model by adding the volumes of all the tetrahedra on the surface.
7. Output wound volume and the wound model.

6.3.3.2 Surface Reconstruction Convex Hull Approximation (Delaunay Tetrahedralization)

A method of constructing solids out of surfaces is to use convex hull approximation (Delaunay tetrahedralization). A tetrahedralization of V is a set T of tetrahedra in 3D whose vertices collectively are V, whose interiors do not intersect each other and whose union is the convex hull of V. A convex hull, $CH(V)$, is the smallest polyhedron in which all elements of V are on or in its interior. The convex hull approximation encloses all the vertices representing

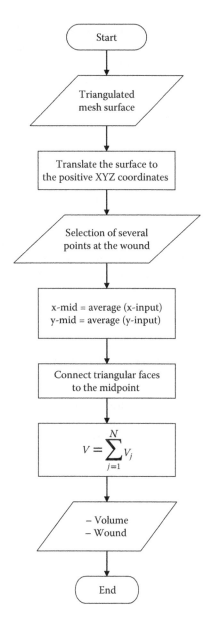

FIGURE 6.20
Flow chart of the midpoint projection algorithm.

the surfaces in the smallest polyhedron. The MATLAB Quickhull algorithm for convex hulls has been used to create the convex hull approximation. The construction of a convex hull out of a 3-simplex (tetrahedron) is performed by growing tetrahedron vertex by vertex, thus constructing more tetrahedrons. A convex hull is constructed by

1. Building a tetrahedron using four points (Delaunay 3-simplex), which is used as a seed upon which the remaining Delaunay tetrahedra crystallize one by one.

2. When adding the next point, the visible faces have to be found. For any integer $r \geq 1$, let $P_r := \{p_1, p_2, ..., p_r\}$. If the new point to be added is inside the current convex hull it will be ignored. If the new point is outside the current hull the visible facets form a connected region on the surface of $CH(P_{r-1})$, called the visible region, or p_r on $CH(P_{r-1})$, which is enclosed by a closed curve consisting of edges $CH(P_{r-1})$. This curve is called the horizon of p_r on $CH(P_{r-1})$ and is shown in Figure 6.21. The horizon of p_r plays a crucial role when transforming $CH(P_{r-1})$ to $CH(P_r)$.

3. Finally, the new point p_r acts as an apex for all the visible faces in its horizon, adding several tetrahedra to the convex hull. While growing the convex hull, a point p_r will be added to the convex hull of P_{r-1}, thus transforming $CH(P_{r-1})$ into $CH(P_r)$, as shown in Figure 6.22.

There are four main algorithms for constructing the convex hull of a set of vertices. The results are identical, with differences in algorithm execution time. The algorithms are incremental with complexity of O $(n \log n)$, gift

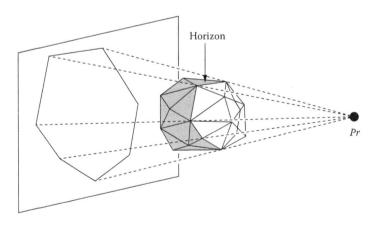

FIGURE 6.21
The horizon of a polytope.

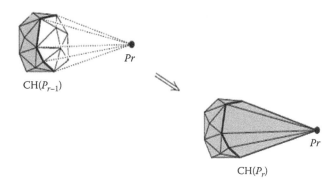

FIGURE 6.22
Adding a point to the convex hull.

wrapping with complexity O (n^2), divide and conquer with complexity of O ($n \log n$), and Quickhull with complexity of O ($n \log n$).

Figure 6.23 shows the results of constructing solids from a model and a real scan using convex hull approximation. Figure 6.23(a) shows a wound model and Figure 6.23(b) shows its reconstructed solid using convex hull approximation. Figure 6.23(c) shows the wound surface scan and Figure 6.23(d) shows its solid model created using convex hull approximation.

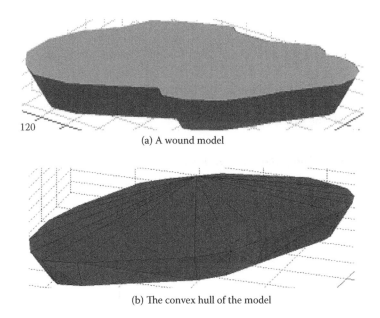

(a) A wound model

(b) The convex hull of the model

FIGURE 6.23
Convex hull approximation for a model and a skin surface: (a) a wound model and (b) its reconstructed solid; (c) the wound surface scan and (d) its solid model. (*Continued*)

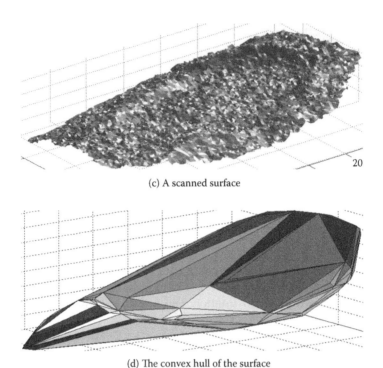

(c) A scanned surface

(d) The convex hull of the surface

FIGURE 6.23
(*Continued*) Convex hull approximation for a model and a skin surface: (a) a wound model and
(b) its reconstructed solid; (c) the wound surface scan and (d) its solid model.

The reconstructed solid is composed of a set of tetrahedra that encloses all
the vertices on the surface. The volume of the solid is equal to the sum of the
volumes of all the tetrahedra.

6.2.3.3 Volume Computation from Reconstructed Solids

In the case of solids consisting of several tetrahedra, the volume of the model can
be obtained by computing the volume of each tetrahedron and totaling the volume for all the tetrahedra. In this way the volume of the wound can be easily computed. A tetrahedron is a polyhedron composed of four triangular faces, three of
which meet at each vertex. For a tetrahedron with vertices $P_1 = (x_1, y_1, z_1)$, $P_2 = (x_2, y_2, z_2)$, $P_3 = (x_3, y_3, z_3)$, and $P_4 = (x_4, y_4, z_4)$, the volume is given by equation (6.1):

$$V_j = \left(\frac{1}{3!}\right) \begin{vmatrix} x_1 & y_1 & z_1 & 1 \\ x_2 & y_2 & z_2 & 1 \\ x_3 & y_3 & z_3 & 1 \\ x_4 & y_4 & z_4 & 1 \end{vmatrix} \tag{6.1}$$

To compute the volume of the solid, the volumes for all the tetrahedra contained in the solid are summed as shown in equation (6.2):

$$V = \sum_{j=1}^{N} V_j, \text{ where N is the number of tetrahedra.} \tag{6.2}$$

6.2.3.4 Conventional Method for Volume Computation of a Leg Ulcer

Volume is the amount of 3D space occupied by an object or geometric and is usually quantified using a volume unit, typically cubic meters (m^3). It can also be numerically quantified as cubic millimeters (mm^3), cubic centimeters (cm^3), or any other relevant unit depending of the size of the object being measured. In the case where the object is not regular and cannot be calculated using an arithmetic formula, volume can be determined using Archimedes' principle.

Archimedes' principle states that an object fully or partially immersed in a fluid is buoyed up by a force equal to the weight of the fluid that the object displaces. There are three volume-measuring techniques that use Archimedes' principle: the level, overflow, and suspension methods. Figure 6.24 shows schematic diagrams of the level, overflow, and suspension methods of measuring volume.

Mold material is a flexible material that can be formed into any shape and its volume can be measured based on Archimedes' principle. Before the measurements are performed, an electronic balance is calibrated in order to obtain reliable and accurate measurements. For the measurements, it is assumed that 1 ml of water equals 1 g and that 1 ml of water equals 1 mm^3.

The weight of an empty measuring cylinder, x_1, is measured and recorded. The measuring cylinder is then filled with water up to 10 ml and the reading is recorded as x_2. The net weights of each ulcer prototype model, $y1$, are being measured. After measuring the net weight of the models and measuring the

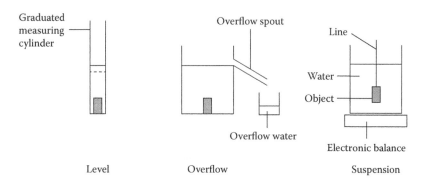

| Level | Overflow | Suspension |

FIGURE 6.24
Schematic diagram of the level, overflow, and suspension methods of measuring volume [73].

cylinder, the models' cavities are filled with mold material, which is weighed and recorded as y_2. The weight of the mold material is calculated by subtracting the weight of the measuring cylinder. The amount of mold material being used is then placed into the measuring cylinder followed by the water. The water is filled to 10 ml. The final reading is x_3. The final reading of the cavity volume for each model is calculated by subtracting x_2 from x_3. Since 1 ml of water is equal to 1 g or 1000 mm^3, the water used to fill up the 10 ml measuring cylinder is equivalent to the cavity volume.

Example:

Model weight, $y1 = 19.002$ g

Model weight + mold material, $y_2 = 21.047$ g

Empty measuring cylinder, $x_1 = 27.012$ g

Measuring cylinder filled with 10 ml of water, $x_2 = 36.794$ g

Measuring cylinder + mold material + water filled up to 10 ml, $x_3 = 37.621$ g

Volume of the model cavity $= x_3 - x_2 = 37.621$ g $- 36.794$ g $= 0.827$ g $= 827$ mm^3

6.3.6 Data Acquisition of a Leg Ulcer Wound

The laser triangulation technique is the most common data acquisition technique used to scan and digitize an object surface. It utilizes distance measurement based on similar triangles and trigonometric functions. The color image produced by the laser scanner is low resolution compared to a digital camera. Among the advantages of laser scanning is having an image that is exactly aligned with the 3D geometry, which simplifies the postprocessing work required for the scanned data [31].

Capturing 3D images of leg ulcers was done using a noncontact 3D laser scanner. Figure 6.25 shows the working principle of a laser scanner based on the laser triangulation technique.

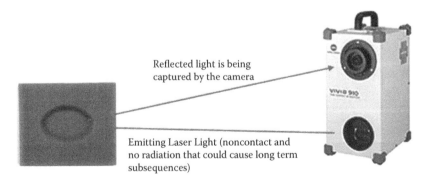

Reflected light is being captured by the camera

Emitting Laser Light (noncontact and no radiation that could cause long term subsequences)

FIGURE 6.25
Working principle of a noncontact 3D laser scanner.

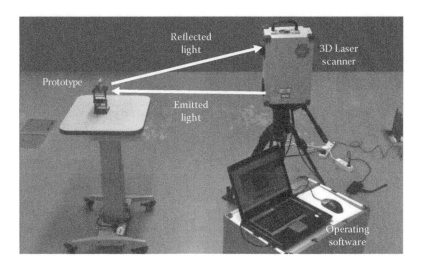

FIGURE 6.26
Setup of the noncontact 3D laser scanner.

Figure 6.26 shows the setup of the laser scanning device. Following the manufacturer's recommendations, the distance from the lens of the laser scanner to the prototype is set between 0.6 m and 1.2 m. The scanned images will be blurry, creating a rough surface, if the object is located further away than 1.2 m. Similarly, when the object is too close to the scanner, no object will be captured.

The preliminary setup and a luminance check are done prior to the surface scan to ensure successful image acquisition. A surface scan of the prototype is then obtained as shown in Figure 6.27.

FIGURE 6.27
Software interface and operating system performing a surface scan.

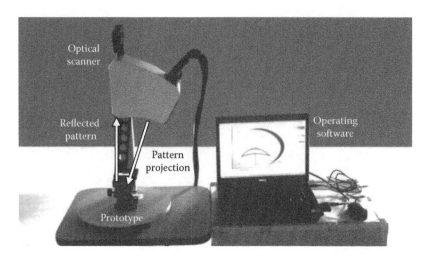

FIGURE 6.28
Setup of an optical scanner.

An optical scanner is another type of scanner that can be used to capture surface information. The setup of an optical scanner and schematic diagram are shown in Figures 6.28 and 6.29. Its working principle is based on the structured light data acquisition technique with phase shifting. Patterns generated from the projector are projected onto the object surface and the reflective patterns are captured with a camera, as illustrated in Figure 6.30.

Figure 6.31 shows the deviations of patterns that represent changes in the object's height. Prior to surface scanning, the object is required to be in full focus, with the crosshairs blended in with the target, as shown in Figure 6.31.

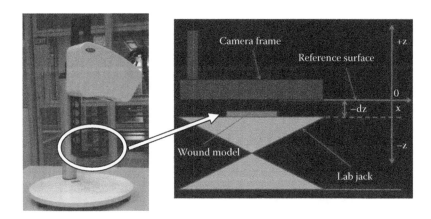

FIGURE 6.29
Schematic of an optical scanner.

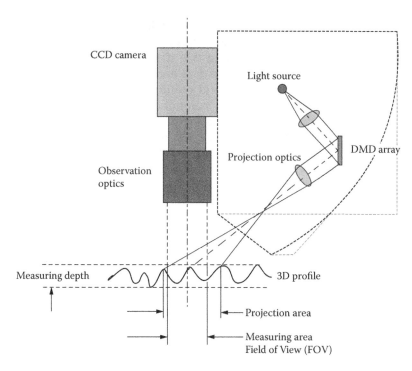

FIGURE 6.30
Working principle of an optical scanner.

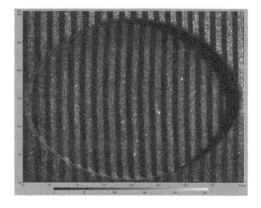

FIGURE 6.31
Fringe pattern on a targeted object.

In this work, an optical scanner is used as additional equipment for data acquisition to facilitate comparative studies between the laser triangulation and structured light methods. These two data acquisition methods were used to scan and digitize the surface data and the outputs are in the form of points of cloud. The points of cloud require data processing, including sampling, aligning, merging multiple points, hole filling, and smoothing, for surface reconstruction prior to volume computation, as shown in Table 6.2.

Prior to computing the volume, a solid model must be reconstructed from the surface scan. Two algorithms—midpoint projection and convex hull approximation—were used to reconstruct the solid model from the cropped ulcer wound cavity. Each algorithm reconstructs the surface to

TABLE 6.2

Process of Data Processing

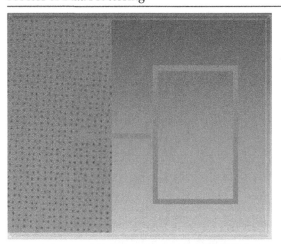	Points of cloud generated by scanning and digitizing the ulcer wound surface.
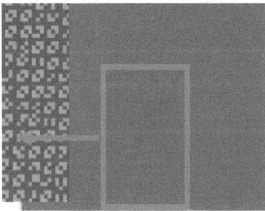	Points of cloud are connected in triangles or quadrilaterals (for two dimensions) or tetrahedra or hexahedral bricks (for three dimensions) that form the triangulation mesh, also known as meshing.

(Continued)

TABLE 6.2 (CONTINUED)

Process of Data Processing

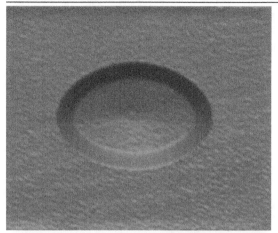

The cavity boundary is selected and cropped out for cavity volume measurement.

Due to the characteristics of the CAD file being in a triangulation mesh form, the cropped boundary indicates an irregular zigzag shape that may introduce some overestimation in volume and thus requires some processing.

(Continued)

TABLE 6.2 (CONTINUED)

Process of Data Processing

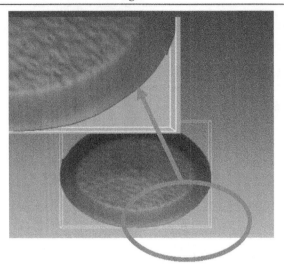

Fit boundary to curve function is used to rearrange or smooth the boundary meshes with curves. This operation is used to repair the boundary edges introduced during the scanning process.

form a solid in a different way. The suitability of each algorithm for different ulcer wound shapes and attributes will be analyzed and discussed. Table 6.3 shows the surface reconstruction using the midpoint projection and convex hull approximation algorithms.

6.3.7 Volume Measurement of the Ulcer Wound Model

There are several factors that will directly affect volume computation accuracy. The first factor that can affect volume computation accuracy is wound attributes. These include wound boundary, wound edges, wound base, and wound depth.

Another aspect that can improve the volume assessment is the data acquisition technique. Each technique has its strengths and limitations in acquiring 3D data. Two 3D acquisition techniques—structured light (optical scanner) and laser triangulation (laser scanner)—are used to acquire the wound surfaces.

The third factor that can affect volume computation accuracy is the volume computation algorithm. Midpoint projection and convex hull approximation are two algorithms that are used to compute the cavity volume. The strengths, limitations, and suitability of each algorithm are discussed to improve volume assessment of the ulcer wound.

The validation of various wound attributes is very important to ensure the entire process and system produces consistent and desirable results. It

TABLE 6.3

Surface Reconstruction and Volume Computation

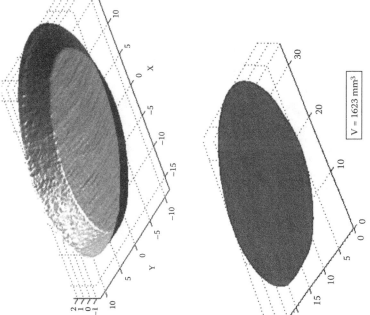

The reconstructed surface for the cropped ulcer cavity model is generated with two selected algorithms (midpoint projection and convex hull approximation).

The midpoint projection algorithm reconstructs the surface by connecting all the points in the model to the calculated midpoint to form a tetrahedral. The volume of the model cavity is calculated by summing the total volume of the tetrahedral generated.

(Continued)

TABLE 6.3 (CONTINUED)

Surface Reconstruction and Volume Computation

(a) Surface reconstruction using the convex hull algorithm is created by connecting the farthest vertices around the surface. The resulting shape will not take account of all vertices in the original shape.

(b) The original model is divided into several divisions as predetermined by the user to give a better approximation in terms of volume. The volume of the model cavity is calculated by summing the total volume of the tetrahedral generated.

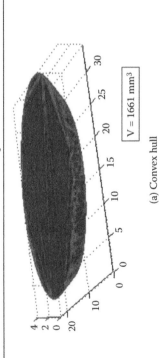

$V = 1661 \text{ mm}^3$

(a) Convex hull

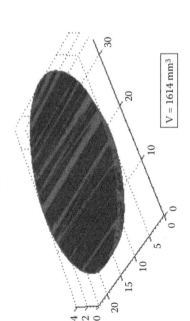

$V = 1614 \text{ mm}^3$

(b) Convex hull with surface division 10x

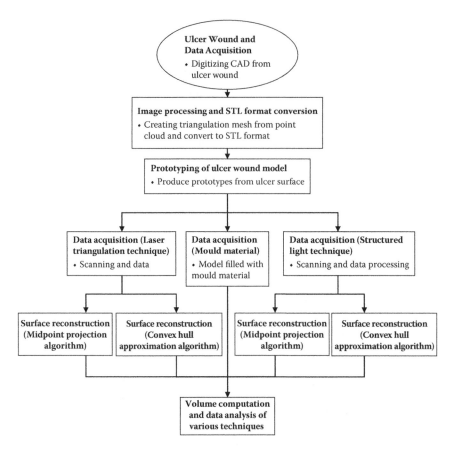

FIGURE 6.32
Volume measurement of the ulcer wound model.

also serves as a guideline to analyze the ulcer wound model. Figure 6.32 illustrates the volume measurement validation and comparison process. As previously discussed, three different approaches can be used for volume measurement: the conventional method using mold material and the reverse engineering methods of laser triangulation and the structured light technique.

In the ulcer wound model, volume is computed using mold material, which serves as the reference volume for the comparison between laser triangulation and the structured light technique. Seventeen various wound attribute models were built based on the recommendations from dermatologists. The volumes of the models are known from the CAD modeling software.

6.4 Results and Analysis

6.4.1 Volume Computation of the Ulcer Wound Model

The volume computation process begins with the model of a known volume followed by the ulcer wound model. Twenty-three patients with consent were involved in this research, providing twenty-six ulcer wound models. The volume computation using mold material gives a higher value of R^2, with an average error of less than 8%. However, it is intrusive and not suitable for use on wounds. Hence it serves as a reference volume for the comparative studies of different data acquisition techniques and algorithms.

In the first stage, volume measurements were performed on the various wound attribute models with known volume. Seventeen models were involved in obtaining the coefficient of determination for each volume computation technique. Conventional cavity volume measurement, weighing the amount of mold material used to fill up the cavity, shows the highest R^2 value, at 0.998818, with an error of less than 8%.

The optical scanning approach provides better accuracy, for both error percentage and R^2, compared with laser scanning. Convex hull approximation provides better accuracy in volume computation for all cases except for the elevated base models.

The second stage shows the results for the volume measurements on the ulcer wound model. Twenty-six ulcer wound prototypes were used. Conventional cavity volume measurements are very accurate and hence served as reference volumes for the comparisons. In most of the models, optical scanning outperforms laser scanning except for the models with an elevated base. Optical scanning gives an R^2 value of 0.999979, whereas laser scanner gives an R^2 value of 0.999976. Volume computation using midpoint projection has better accuracy in most wound models, while convex hull approximation gives better results for various wound attribute models that are less complex in the overall shape.

R^2 is used to help in predicting the experimental results when the reference volume is given. The reference volume used is based on the volume obtained from volume computation using mold material. It shows that the volume computation result obtained from an optical scanner gives the best fitting results, followed by the laser scanner.

6.4.2 Comparison of the Laser Triangulation and Structured Light Data Acquisition Techniques Using the Midpoint Projection Algorithm

Based on Figure 6.33, the error percentage for the twenty-six wound models obtained from optical scanning is always smaller than that for laser scanning except for ulcer wound model 14.

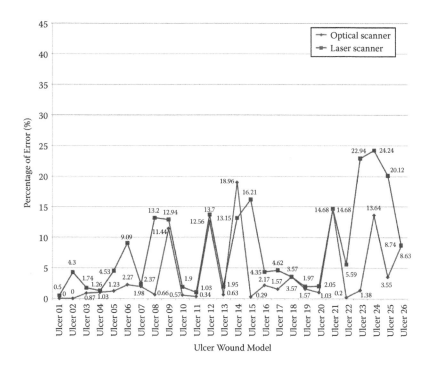

FIGURE 6.33
Error percentage for models using midpoint projection.

6.4.3 Comparison of the Laser Triangulation and Structured Light Data Acquisition Techniques Using the Convex Hull Approximation Algorithm

The R^2 values for optical and laser scanning are 0.9999767 and 0.9999766, respectively. Figure 6.34 shows the error percentage for the 26 wound models using the convex hull algorithm. Among the ulcer models, there are seven models generated from optical scanning that have higher errors than for laser scanning.

Seven ulcer models (7, 10, 12, 14, 15, 22, and 26) show a different trend compared with the others; that is, convex hull approximation with optical scanning is better than laser scanning. Ulcers 12, 14, and 15 have similar wound bases, which is an elevated base. It is clear that convex hull approximation is not suitable to compute the volume a wound cavity that has an elevated base at the wound bed.

For ulcers 7, 10, and 26, the 3D scanned surface shows some deviation that affects the accuracy of the volume computation. The deviation error occurred due to scanning limitations of the optical scanner to rapid changes in the height of the wound bed, as shown in Figure 6.35.

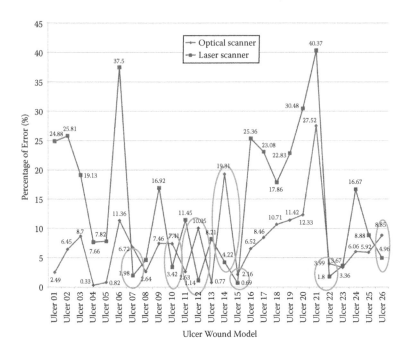

FIGURE 6.34
Error percentage for models using convex hull approximation.

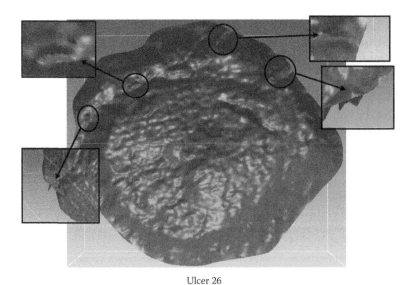

Ulcer 26

FIGURE 6.35
Deviation of the 3D surface scan of the wound model.

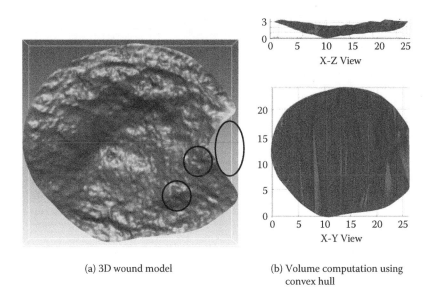

(a) 3D wound model

(b) Volume computation using convex hull

FIGURE 6.36
Errors in ulcer 22.

For ulcer 22, some of the areas are elevated above the reconstructed surface, thus increasing the volume of the model, leading to overestimation for the model. In addition, the irregular shape of the boundary caused overestimation in the volume measurement, as shown in Figure 6.36. However, most of the 3D wound models obtained from the optical scanner have better accuracy in volume computation using convex hull approximation.

6.4.4 Comparison of Midpoint Projection and the Convex Hull Approximation Algorithm for Each Scanning Technique

There were twenty-six 3D scanned wound models obtained from each scanning method. Referring to Figure 6.37, ulcer models 4, 5, 9, 12, and 24 show higher errors when the cavity volume is computed using midpoint projection. In contrast, the other ulcer models showed higher errors when the cavity volume was computed using convex hull approximation. The same phenomenon occurs when the 3D scanned surface is obtained using the laser scanner.

Figure 6.38 shows that ulcer models 7, 8, 12, 14, 15, and 22 have higher errors when the volume cavity is computed using midpoint projection. Both laser and optical scanners show the same trend whereby better accuracy is obtained when the volume is computed using midpoint projection. The convex hull algorithm shows higher accuracy for most of the ulcer wound

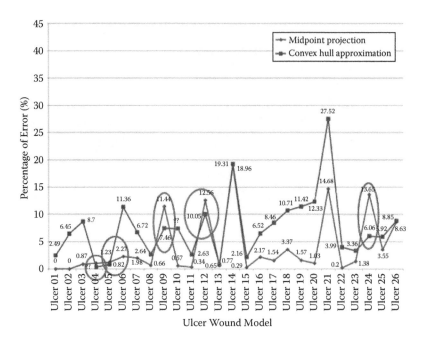

FIGURE 6.37
Optical scanner with midpoint projection and convex hull approximation algorithm.

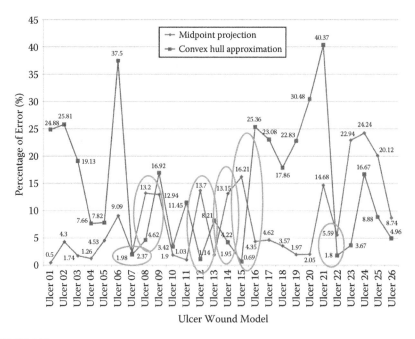

FIGURE 6.38
Laser scanner with midpoint projection and convex hull approximation algorithm.

models. As the surface division increases, the model becomes more like the original shape of the model. However, convex hull approximation tends to introduce more gaps as the surface division increases.

In the study on different data acquisition techniques and volume computation algorithms using different wound attribute models with known volume, convex hull approximation has better accuracy compared with midpoint projection. Convex hull approximation is very much dependent on the points for the 3D model. If the model boundary is more complex, more surface division is needed, and the gap introduced is significant. For models with less complexity in boundary shape, less surface division is needed and hence the gaps introduced are less significant.

6.5 Conclusion

The effectiveness of a treatment regime can be estimated by measuring changes in the ulcer wound. The assessment of wounds covers a range of observations to determine the wound status at different times. Wound measurements that include top area, true surface area, depth, and volume represent objective parameters that can be used to determine the progression or regression of a wound.

Invasive methods for wound measurements are time consuming and often result in inconsistency in patient care. When computing wound area, overestimation occurs due to the fixed size of the squares composing the grid used for calculation. This overestimation is solved by digitizing the wound boundary and dividing the area into smaller units for better accuracy. Overestimation in the volume calculation is caused by overfilling the wound with material, such as saline or alginate paste. In addition, invasive methods for volume estimation might cause infection or disturbance of the wound tissue and therefore are limited in practical use. Invasive methods for measuring volume are not suitable for large wounds, wounds that lay on a limb with high curvature, and shallow wounds.

Noninvasive methods for wound measurements from 3D surface scans will help in obtaining measurements that are more objective and eliminate the problems associated with the invasive methods. Current noninvasive methods for volume computation suffer from limitations such as overestimation due to the existence of irregularities in the wound surface, large and high-curvature areas, and dealing with depressed edges wounds.

Noninvasive computer algorithms that objectively determine ulcer wound parameters such as top area, true surface area, average depth, and volume has been developed. The algorithms are based on 3D surface scans and the wound of interest has been limited to leg ulcers. The methods overcome the problems faced by invasive methods.

A 3D laser scanner is used to scan the wound surface using the triangulation concept, in which the depth of a large number of points located at the scanned surface is accurately obtained. In this chapter, algorithms for measuring wound top area, true surface area, average depth, and volume from 3D surface scans were developed and investigated.

Wound attributes that describe the wounds are modeled on real ulcer wound surface images. Wound models representing possible ulcer wounds are created using CAD software. These models are used to investigate the performance of solid reconstruction methods. In addition, molded surfaces of the wounds are created using rapid prototyping, to verify the wound parameters (top area, true surface area, average depth, and volume) measured using invasive and noninvasive methods.

In this work, midpoint projection was used to reconstruct solids from surfaces. In addition, convex hull approximation was investigated. These methods do not require a large area around the wound for surface interpolation, thus these methods are not affected by irregularities (swelling and scales) in the surrounding skin surface. The volume is computed from scattered data to handle depressed edges.

The study revealed that both the structured light (optical scanner) and laser triangulation (laser scanner) data acquisition techniques can be used for 3D geometrical CAD of the ulcer wound. In generating the CAD model for both data acquisition techniques, when the boundary edge and the elevated base increase to certain level, shadows will appear that affect the accuracy. It was established and verified that the structured light–based 3D technique produces better accuracy compared with the laser triangulation data acquisition technique.

The suitability and reliability of the algorithm for surface reconstruction and volume computation of an ulcer cavity was also determined compared with the conventional method (mold material) of volume cavity measurement. It was shown that both midpoint projection and the convex hull approximation algorithm work on most of the wound attributes. Midpoint projection volume computation is suitable for all wound types, whereas convex hull approximation is suitable for all types of wounds except those with an elevated base. It was found that midpoint projection produces better results in volume measurement and is suitable for all wound attributes introduced.

Volume computation using conventional methods (mold material) showed the highest accuracy among all the measuring methods. However, this method is invasive and can cause discomfort and possible infection and is only suitable for research purposes. This research established that ulcer wounds can be assigned with various wound attributes (boundary, edge, base, and depth). It was also shown that each attribute has an effect on data acquisition and volume computation accuracy. The above work identifies the strengths and limitations of each data acquisition technique and choice of algorithm that can be used to develop a biomedical measuring system—a useful adjunct to conventional clinical evaluation.

Appendix: MATLAB Code

```
% Read the 3D CAD data
%[x2,y2] = cad2matdemo2('C:\Documents and Settings\user\My Documents\mat\
%p12.stl');
[x2,y2] = cad('D:\phd\stl files\tahseen1a.stl');

% Mid point Projection
%[x2,y2] = cad2matdemo2('C:\Documents and Settings\user\My Documents\mat\
%p12.stl');
%[x2,y2] = cad2matdemo2('D:\phd\stl files\tahseen1a.stl');
[x2,y2]  =  cad('C:\Users\USER\Desktop\matlab  codes\scans\optical  scan
cropped\sample1scn1cr1.stl');
%display(x2);
%display(y2);
figure;
%ankur = size(x2);
%display(ankur);
[c cc] = size(y2);
%ankur = size(cc);
%display(ankur);

% align the surface to the positive x,y,z plane by if the min of x,y,z is
% negative by subtracting min-1 from all the axes
% checking the the location of the minimum x,y,z dimensions
n1 = min(y2(:,1));
n2 = min(y2(:,2));
n3 = min(y2(:,3));

% if any of the dimensions lay in the negative axis we will performe
% translation for the whole object or surface
if n1<0
y2(:,1) = y2(:,1)-n1;
end

if n2<0
y2(:,2) = y2(:,2)-n2;
end

if n3<0
y2(:,3) = y2(:,3)-n3;
end

% calculate the mid point
% the mid point lay in the center of the x, y of the object and
% having the
% maximum height.

dif1 = (max(y2(:,1))-min(y2(:,1)))/2;
p(1) = min(y2(:,1))+dif1;
dif2 = (max(y2(:,2))-min(y2(:,2)))/2;
p(2) = min(y2(:,2))+dif2;
p(3) = max(y2(:,3));
% add the origin point as the 4th column in the vertices list to give the
% list of tetrahedrons

y2(c+1,:) = p;

x2(:,4) = c+1;
```

```
x2 = uint32(x2);

display(x2);
simp _ plot _ 3dd(y2,x2);

figure;
vertices = size(y2);
numberoftriangles = size(x2);
display(vertices);
display(numberoftriangles);
vo = vol51(y2,x2);
[k, v2] = convhulln(y2);
trisurf(k,y2(:,1),y2(:,2),y2(:,3));

% global B T;
% T = 1;
% select3dtool
%Remove the repeated vertices in the triangular mesh and rearrange the
%%allocation table...
[B,ix,jx] = unique(y2,'rows');
%[B,I,J] = UNIQUE(...) also returns index vectors I and J such that B = A(I)
% and A = B(J) (or B = A(I,:) and A = B(J,:)).

[a,b] = size(y2);
c = 1;
fori = 1:a/3
e1(i,1) = jx(c);
e1(i,2) = jx(c+1);
e1(i,3) = jx(c+2);
    c = c+3;
end

% The following function will return the vertices that belong to the surface
% boundary (edge boundary)
e = boundedges(B,e1);
figure, simp _ plot _ 2d(B, e);

% Re arrange the input matrix
[a,b] = size(e);
c = 1;
c2 = 1;
fori = 1:a
v01(i,:) = B(e(i,1),:);
v02(i,:) = B(e(i,2),:);
XX(c2) = v01(i,1);
XX(c2+1) = v02(i,1);
YY(c2) = v01(i,2);
YY(c2+1) = v02(i,2);
    c2 = c2+2;
end
% compute the area of the polygon defined by the edge vertices
A = polyarea(XX,YY);
volumeofimage = vo;
volumeofimage = uint32(volumeofimage);
display(volumeofimage);

% Convex hull/Delaunay Tetrahedralization
%[x2,y2] = cad2matdemo2('C:\Documents and Settings\user\My Documents\mat\
%p12.stl');
```

```
[x2,y2] = cad('C:\Users\USER\Desktop\matlab codes\segmented.jpg');
% figure;
[c cc] = size(y2);

 [vol,s,Tes] = div _ 20(y2);
% divide the surface to equally spaced subsurfaces using the Y
% coordiante...
% in this file the surface is divided to 20 surfaces

yy = unique(y2,'rows');
 % if any of the dimensions lay in the negative axis we will performe
% translation for the whole object or surface

[a b] = size(yy);

n1 = min(yy(:,1));
n2 = min(yy(:,2));
n3 = min(yy(:,3));

% if any of the dimensions lay in the negative axis we will performe
% translation for the whole object or surface
if n1<0
yy(:,1) = yy(:,1)-n1;
end

if n2<0
yy(:,2) = yy(:,2)-n2;
end

if n3<0
yy(:,3) = yy(:,3)-n3;
end

cs1 = 1;
cs2 = 1;
cs3 = 1;
cs4 = 1;
cs5 = 1;
cs6 = 1;
cs7 = 1;
cs8 = 1;
cs9 = 1;
cs10 = 1;
cs11 = 1;
cs12 = 1;
cs13 = 1;
cs14 = 1;
cs15 = 1;
cs16 = 1;
cs17 = 1;
cs18 = 1;
cs19 = 1;
cs20 = 1;
dif = max(yy(:,2))-min(yy(:,2));
dis = dif/20;

for in = 1 :a
ifyy(in,2) < min(yy(:,2))+dis
s1(cs1,:) = yy(in,:);
cs1 = cs1+1;
```

```
elseifyy(in,2) < min(yy(:,2))+dis*2
s2(cs2,:) = yy(in,:);
        cs2 = cs2+1;
elseifyy(in,2) < min(yy(:,2))+dis*3
s3(cs3,:) = yy(in,:);
        cs3 = cs3+1;
elseifyy(in,2) < min(yy(:,2))+dis*4
s4(cs4,:) = yy(in,:);
        cs4 = cs4+1;
elseifyy(in,2) < min(yy(:,2))+dis*5
s5(cs5,:) = yy(in,:);
        cs5 = cs5+1;

elseifyy(in,2) < min(yy(:,2))+dis*6
s6(cs6,:) = yy(in,:);
cs6 = cs6+1;
elseifyy(in,2) < min(yy(:,2))+dis*7
s7(cs7,:) = yy(in,:);
        cs7 = cs7+1;
elseifyy(in,2) < min(yy(:,2))+dis*8
s8(cs8,:) = yy(in,:);
                cs8 = cs8+1;
elseifyy(in,2) < min(yy(:,2))+dis*9
s9(cs9,:) = yy(in,:);
                cs9 = cs9+1;
elseifyy(in,2) < min(yy(:,2))+dis*10
s10(cs10,:) = yy(in,:);
                cs10 = cs10+1;
elseifyy(in,2) < min(yy(:,2))+dis*11
s11(cs11,:) = yy(in,:);
cs11 = cs11+1;
elseifyy(in,2) < min(yy(:,2))+dis*12
s12(cs12,:) = yy(in,:);
            cs12 = cs12+1;
elseifyy(in,2) < min(yy(:,2))+dis*13
s13(cs13,:) = yy(in,:);
                cs13 = cs13+1;
elseifyy(in,2) < min(yy(:,2))+dis*14
s14(cs14,:) = yy(in,:);
                cs14 = cs14+1;
elseifyy(in,2) < min(yy(:,2))+dis*15
s15(cs15,:) = yy(in,:);
                cs15 = cs15+1;
elseifyy(in,2) < min(yy(:,2))+dis*16
s16(cs16,:) = yy(in,:);
cs16 = cs16+1;
elseifyy(in,2) < min(yy(:,2))+dis*17
s17(cs17,:) = yy(in,:);
        cs17 = cs17+1;
elseifyy(in,2) < min(yy(:,2))+dis*18
s18(cs18,:) = yy(in,:);
                cs18 = cs18+1;
elseifyy(in,2) < min(yy(:,2))+dis*19
s19(cs19,:) = yy(in,:);
                cs19 = cs19+1;
else
s20(cs20,:) = yy(in,:);
                cs20 = cs20+1;

end
```

```
end

vol = 0;
si = 0;
if cs1>3
Tes1 = delaunay3(s1(:,1),s1(:,2),s1(:,3), {'Qt', 'Qbb', 'Qc', 'Qz'});
Tes = [Tes1];
v1 = vol51(s1, Tes1);
vol = vol+v1;
[si1 b] = size(s1);
si = si+si1;
s = s1;
end

if cs2>3
Tes2 = delaunay3(s2(:,1),s2(:,2),s2(:,3), {'Qt', 'Qbb', 'Qc', 'Qz'});
v2 = vol51(s2, Tes2);
vol = vol+v2;
[si2 b] = size(s2);
Tes = [Tes;Tes2+si];
si = si+si2;
s = [s;s2];
end

if cs3>3
Tes3 = delaunay3(s3(:,1),s3(:,2),s3(:,3), {'Qt', 'Qbb', 'Qc', 'Qz'});
v3 = vol51(s3, Tes3);
vol = vol+v3;
[si3 b] = size(s3);
Tes = [Tes;Tes3+si];
si = si+si3;
s = [s;s3];

end

if cs4>3
Tes4 = delaunay3(s4(:,1),s4(:,2),s4(:,3), {'Qt', 'Qbb', 'Qc', 'Qz'});
v4 = vol51(s4, Tes4);
vol = vol+v4;
[si4 b] = size(s4);
Tes = [Tes;Tes4+si];
si = si+si4;
s = [s;s4];

end

if cs5>3
Tes5 = delaunay3(s5(:,1),s5(:,2),s5(:,3), {'Qt', 'Qbb', 'Qc', 'Qz'});
v5 = vol51(s5, Tes5);
vol = vol+v5;
[si5 b] = size(s5);
Tes = [Tes;Tes5+si];
si = si+si5;
s = [s;s5];

end

if cs6>3
Tes6 = delaunay3(s6(:,1),s6(:,2),s6(:,3), {'Qt', 'Qbb', 'Qc', 'Qz'});
v6 = vol51(s6, Tes6);
vol = vol+v6;
```

```
[si6 b] = size(s6);
Tes = [Tes;Tes6+si];
si = si+si6;
s = [s;s6];

end

if cs7>3
Tes7 = delaunay3(s7(:,1),s7(:,2),s7(:,3), {'Qt', 'Qbb', 'Qc', 'Qz'});
v7 = vol51(s7, Tes7);
vol = vol+v7;
[si7 b] = size(s7);
Tes = [Tes;Tes7+si];
si = si+si7;
s = [s;s7];
end

if cs8>3
Tes8 = delaunay3(s8(:,1),s8(:,2),s8(:,3), {'Qt', 'Qbb', 'Qc', 'Qz'});
v8 = vol51(s8, Tes8);
vol = vol+v8;
[si8 b] = size(s8);
Tes = [Tes;Tes8+si];
si = si+si8;
s = [s;s8];

end

if cs9>3
Tes9 = delaunay3(s9(:,1),s9(:,2),s9(:,3), {'Qt', 'Qbb', 'Qc', 'Qz'});
v9 = vol51(s9, Tes9);
vol = vol+v9;
[si9 b] = size(s9);
Tes = [Tes;Tes9+si];
si = si+si9;
s = [s;s9];

end

if cs10>3
Tes10 = delaunay3(s10(:,1),s10(:,2),s10(:,3), {'Qt', 'Qbb', 'Qc', 'Qz'});
v10 = vol51(s10, Tes10);
vol = vol+v10;
[si10 b] = size(s10);
Tes = [Tes;Tes10+si];
si = si+si10;
s = [s;s10];

end

if cs11>3
Tes11 = delaunay3(s11(:,1),s11(:,2),s11(:,3), {'Qt', 'Qbb', 'Qc', 'Qz'});
v11 = vol51(s11, Tes11);
vol = vol+v11;
[si11 b] = size(s11);
Tes = [Tes;Tes11+si];
si = si+si11;
s = [s;s11];

end
```

```
if cs12>3
Tes12 = delaunay3(s12(:,1),s12(:,2),s12(:,3), {'Qt', 'Qbb', 'Qc', 'Qz'});
v12 = vol51(s12, Tes12);
vol = vol+v12;
[si12 b] = size(s12);
Tes = [Tes;Tes12+si];
si = si+si12;
s = [s;s12];

end

if cs13>3
Tes13 = delaunay3(s13(:,1),s13(:,2),s13(:,3), {'Qt', 'Qbb', 'Qc', 'Qz'});
v13 = vol51(s13, Tes13);
vol = vol+v13;
[si13 b] = size(s13);
Tes = [Tes;Tes13+si];
si = si+si13;
s = [s;s13];
end

if cs14>3
Tes14 = delaunay3(s14(:,1),s14(:,2),s14(:,3), {'Qt', 'Qbb', 'Qc', 'Qz'});
v14 = vol51(s14, Tes14);
vol = vol+v14;
[si14 b] = size(s14);
Tes = [Tes;Tes14+si];
si = si+si14;
s = [s;s14];
end

if cs15>3
Tes15 = delaunay3(s15(:,1),s15(:,2),s15(:,3), {'Qt', 'Qbb', 'Qc', 'Qz'});
v15 = vol51(s15, Tes15);
vol = vol+v15;
[si15 b] = size(s15);
Tes = [Tes;Tes15+si];
si = si+si15;
s = [s;s15];
end

if cs16>3
Tes16 = delaunay3(s16(:,1),s16(:,2),s16(:,3), {'Qt', 'Qbb', 'Qc', 'Qz'});
v16 = vol51(s16, Tes16);
vol = vol+v16;
[si16 b] = size(s16);
Tes = [Tes;Tes16+si];
si = si+si16;
s = [s;s16];
end

if cs17>3
Tes17 = delaunay3(s17(:,1),s17(:,2),s17(:,3), {'Qt', 'Qbb', 'Qc', 'Qz'});
v17 = vol51(s17, Tes17);
vol = vol+v17;
[si17 b] = size(s17);
Tes = [Tes;Tes17+si];
si = si+si17;
s = [s;s17];
end
```

```
if cs18>3
Tes18 = delaunay3(s18(:,1),s18(:,2),s18(:,3), {'Qt', 'Qbb', 'Qc', 'Qz'});
v18 = vol51(s18, Tes18);
vol = vol+v18;
[si18 b] = size(s18);
Tes = [Tes;Tes18+si];
si = si+si18;
s = [s;s18];
end

if cs19>3
Tes19 = delaunay3(s19(:,1),s19(:,2),s19(:,3), {'Qt', 'Qbb', 'Qc', 'Qz'});
v19 = vol51(s19, Tes19);
vol = vol+v19;
[si19 b] = size(s19);
Tes = [Tes;Tes19+si];
si = si+si19;
s = [s;s19];
end

if cs20>3
Tes20 = delaunay3(s20(:,1),s20(:,2),s20(:,3), {'Qt', 'Qbb', 'Qc', 'Qz'});
v20 = vol51(s20, Tes20);
vol = vol+v20;
Tes = [Tes;Tes20+si];
s = [s;s20];
end

vol = uint32(vol)
display(vol);

figure, simp_plot_3dd(s,Tes);
```

References

1. MAVIS II: 3D Wound Instrument Measurement. https://www.comp.glam. ac.uk/pages/staff/pplassma/Medimaging/Projects/Wounds/Mavis-ii/ Index.html
2. Callieri M, Cignoni P, Pingi P, Scopigno R, Coluccia M, Gaggio G, Romanelli M. Derma: monitoring the evolution of skin lesions with a 3D system. Paper presented at the Vision, Modeling, and Visualization Conference 2003, Munich, Germany.
3. Werdin F, Tennenhaus M, Schaller HE, Rennekampff HO. Evidence-based management strategies for treatment of chronic wounds. *Eplasty* 2009;9:e19.
4. Wolff K, Fitzpatrick TB, Johnson RA. *Fitzpatrick's color atlas and synopsis of clinical dermatology.* New York: McGraw-Hill, 2009.
5. Woodbury MG, Houghton PE, Campbell KE, Keast DH. Development, validity, reliability, and responsiveness of a new leg ulcer measurement tool. *Advances in Skin and Wound Care* 2004;17:187–196.
6. Kecelj-Leskovec N, Jezeršek M, Možina J, Pavlović MD, Lunder T. Measurement of venous leg ulcers with a laser-based three-dimensional method: comparison to computer planimetry with photography. *Wound Repair and Regeneration* 2007;15:767–771.

7. Wild T, Prinz M, Fortner N, Krois W, Sahora K, Stremitzer S, Hoelzenbein T. Digital measurement and analysis of wounds based on colour segmentation. *European Surgery* 2008;40:5–10.
8. Shai A, Maibach HI. Ulcer measurement and patient assessment. In *Wound healing and ulcers of the skin: diagnosis and therapy—the practical approach.* Berlin: Springer, 2005:89–102.
9. Kundin JI. Apparatus and method for measuring deformed areas of skin surface. U.S. Patent 4483075, November 20, 1984.
10. Goldman RJ, Salcido R. More than one way to measure a wound: an overview of tools and techniques. *Advances in Skin and Wound Care* 2002;15:236–243.
11. Langemo D, Anderson J, Hanson D, Hunter S, Thompson P. Measuring wound length, width, and area: which technique? *Advances in Skin and Wound Care* 2008;21:42–45.
12. Shai A, Maibach HI. Wound healing and ulcers of the skin: diagnosis and therapy—the practical approach. http://dx.doi.org/10.1007/b138035
13. Baranoski S, Ayello EA. *Wound care essentials: practice principles.* Philadelphia: Lippincott Williams & Wilkins, 2007.
14. de la Brassinne M, Thirion L, Horvat LI. A novel method of comparing the healing properties of two hydrogels in chronic leg ulcers. *Journal of the European Academy of Dermatology and Venereology* 2006; 20:131–135.
15. Herbin M, Bon FX, Venot A, Jeanlouis F, Dubertret ML, Dubertret L, Strauch G. Assessment of healing kinetics through true color image processing. *IEEE Transactions on Medical Imaging* 1993;12:39–43.
16. Mekkes JR, Westerhof W. Image processing in the study of wound healing. *Clinics in Dermatology* 1995;13:401–407.
17. Smith RB, Rogers B, Tolstykh GP, Walsh NE, Davis MG, Bunegin L, Williams RL. Three-dimensional laser imaging system for measuring wound geometry. *Lasers in Surgery and Medicine* 1998;23:87–93.
18. Chen F, Brown GM, Song M. Overview of three-dimensional shape measurement using optical methods. *Optical Engineering* 2000;39:10–22.
19. Ozturk C, Dubin S, Schafer ME, Wen-Yao S, Min-Chih C. A new structured light method for 3-D wound measurement. *Proceedings of the 1996 IEEE Twenty-Second Annual Northeast Bioengineering Conference.* Washington, DC; IEEE, 1996:70–71.
20. Krouskop TA, Baker R, Wilson MS. A noncontact wound measurement system. *Journal of Rehabilitation Research and Development* 2002;39:337–346.
21. Malian A, Azizi A, van den Heuvel FA, Zolfaghari M. Development of a robust photogrammetric metrology system for monitoring the healing of bedsores. Photogrammetric Record 2005;20:241–273.
22. Boersma SM. Photogrammetric wound measurement with a three-camera vision system. *International Archives of Photogrammetry and Remote Sensing* 2000;33:84–91.
23. Plassmann P, Jones TD. MAVIS: a non-intrusive instrument to measure area and volume of wounds. *Medical Engineering and Physics* 1998;20:332–338.
24. Drap P, Sgrenzaroli M, Canciani M, Cannata G, Seinturier J. Laser scanning and close range photogrammetry: towards a single measuring tool dedicated to architecture and archaeology. *Proceedings of the CIPA XIXth International Symposium, Antalya, Turkey.* Istanbul: CIPA, 2003.

25. Malian A, Azizi A, van den Heuvel FA. MEDPHOS: a new photogrammetric system for medical measurement. *International Archives of Photogrammetry Remote Sensing and Spatial Information Sciences* 2004;35:311–316.
26. Malian A, van den Heuvel FA, Azizi A. A robust photogrammetric system for wound measurement. *International Archives of Photogrammetry Remote Sensing and Spatial Information Sciences* 2002;34:264–269.
27. Wannous H, Lucas Y, Treuillet S, Albouy B. A complete 3D wound assessment tool for accurate tissue classification and measurement. *15th IEEE International Conference on Image Processing*. Washington, DC: IEEE, 2008:2928–2931.
28. Wannous H, Treuillet S, Lucas Y. Supervised tissue classification from color images for a complete wound assessment tool. *29th Annual International Conference of the IEEE Engineering in Medicine and Biology Society*. Washington, DC: IEEE, 2007:6031–6034.
29. Wannous H, Treuillet S, Lucas Y, Albouy B. Mapping classification results on 3D model: a solution for measuring the real areas covered by skin wound tissues. *3rd International Conference on Information and Communication Technologies: From Theory to Applications*. Washington, DC: IEEE, 2008:1–6.
30. Hani A, Eltegani N, Hussein S, Jamil A, Gill P. Assessment of ulcer wounds size using 3D skin surface imaging. In Badioze Zaman H, Robinson P, Petrou M, Olivier P, Schröder H, Shih T, eds., *Visual informatics: bridging research and practice*, vol. 5857. Berlin: Springer, 2009:243–253.
31. Rocchini C, Cignoni P, Montani C, Scopigno R. Acquiring, stitching and blending diffuse appearance attributes on 3D models. *Visual Computer* 2002;18:186–204.

7

Grading of Acne Vulgaris Lesions

Aamir Saeed Malik, Jawad Humayun, Felix
Boon-Bin Yap, and Javed Khan

CONTENTS

7.1 Introduction to Acne

Acne is one of the most common skin-related diseases and approximately 85% of the teenage population suffers from it [1]. Acne lesions appear on those skin regions where the sebaceous follicles occur in large numbers, including the face, chest, and back [2]. The prevalence of acne is 92% for the face, 45% for the chest, and 61% for the back [3]. According to a survey conducted in United States, more than US$100 million is spent on acne treatment and therapy every year [4]. In Malaysia, a single acne patient spends approximately RM250 (US$80) , each year on medication and treatment.

Acne lesions appear as a result of follicular blockage due to an excessive increase in sebum secretions. Sebaceous glands located under the skin are connected to tiny pores through sebaceous follicles. These sebaceous glands produce a fluid-like substance called sebum that usually flows through the follicle out of the dermis. This oily fluid can sometimes gets stuck inside the skin pores as a result of follicular blockage. This sebum gets infected with bacteria (*Propinibacterium*), resulting in inflammation and pimples [5]. Figure 7.1 shows the structure of sebaceous glands and hair follicles.

7.1.1 Types of Acne

Based on the pathology of the acne lesion, acne is divided into different types, including cosmetica, medicamentosa, conglobata, excoriee, fulminans,

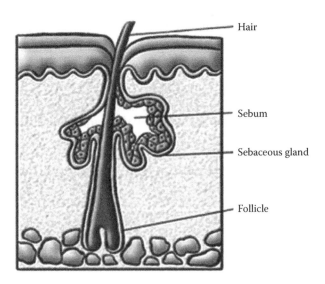

FIGURE 7.1
Structure of the sebaceous glands and hair follicles.

pomade, nuchae, chloracne, keloidalis, mechanica, and rosacea [2]. However, acne vulgaris is the most common type of acne and has been found to constitute 99% of all acne cases. Acne is characterized by factors such as the lesion type and the underlying pathological process. For example, improper and excessive use of cosmetic agents can result in acne cosmetica, while medicamentosa is caused by topical medication applied to the skin; similarly, the main cause of pomade acne is the use of talcum powder. In this chapter we focus on acne vulgaris because of its prevalence. Acne vulgaris lesions are classified into comedones, which are further subcategorized into whitehead and blackhead papules, pustules, nodules, cysts, and scarring [6].

7.1.2 Types of Acne Vulgaris Lesions

Acne vulgaris includes lesions causing no inflammation, like blackhead and whitehead comedones, and inflammatory lesions like papules, pustules, nodules, and cysts. Comedones are the mildest type of acne and are further divided into two types: open comedones or blackheads and closed comedones or whiteheads. A closed comedone, or whitehead, occurs when the sebum completely covers the entrance of the hair follicle and thus no oxidation occurs. Closed comedones appear in the form of little white or skin-colored pimples on the skin surface, as shown in Figure 7.2(a). Open comedones, or blackheads, appear when the accumulation of dead skin cells and the oily substance is exposed to air; the oxidation process occurs and causes the material to turn black, as shown in Figure 7.2(b).

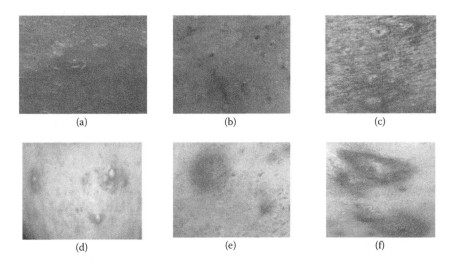

FIGURE 7.2
Types of acne vulgaris: (a) whitehead comedone; (b) blackhead comedone; (c) papule; (d) pustule; (e) nodule; (f) cyst.

In inflammatory acne lesions, the main cause of inflammation, pain, and erythema is infection by *Propinibacterium* when the debris of dead skin cells and oil accumulate in the dermis. Papules are mostly red- or pink-colored raised bumps without any fluidlike substance inside, as in Figure 7.2(c). Papules are large in size compared with comedones and cause inflammation. Pustules are an infected form of a papule. Pustules contain either a yellowish or whitish fluidlike substance called pus and can be surrounded by an area of red color, as shown in Figure 7.2(d). Pus consists of dead skin cells, sebum secretions, and white blood cells and shows signs of infection. Papules and pustules are the most common types of acne vulgaris and constitute about 90% of total cases.

Nodules are red-colored bumps like papules. However, nodules are larger in size and deeper in the skin than papules (Figure 7.2(e)). Nodules occur when the sheath of hair follicles gets torn and the oily substance comes out into the dermis. Nodules are inflammatory and painful lesions. Cysts are mostly red-colored pus-filled lesions (Figure 7.2(f)). Cysts are larger in size than any other acne vulgaris lesion and are present in the lower layer of the skin. Cysts are inflammatory and occasionally painful lesions. When comedones spill over into the surrounding skin area, the local immune system responds to this by producing a puslike liquid. This phenomenon causes cysts. Cysts often leave permanent scars on the skin. Most patients suffer from both inflammatory and noninflammatory acne.

7.1.3 Causes of Acne

The main causes of acne include genetic, hormonal changes, sebaceous activity, bacterial infection, climatic effects, chemical factors, and psychological factors. Thus acne is a multifactorial skin disease, however, genetics is a relatively common reason. Three of four children will suffer from acne if both their parents suffer from acne. But if only one of the two parents has acne, one of four will suffer from acne. Like other genetic diseases, it is possible that acne may not transfer from parents to children. Changes in hormone levels effect the development of acne. Both menstrual cycles and puberty cause such changes in hormone levels. During puberty, the male sex hormone androgen increases, which in turn causes the sebaceous glands, located around hair follicles, to enlarge and produce more sebum. Hyperactivation has an effect on the sebaceous glands. *Propionibacterium*, which is also called *P-acnes*, infects the sebum that is stuck under the skin surface. In hot climates, the sebaceous glands become hyperactive and excrete more sebum. The use of different chemical agents, such as facial washes, is a main factor causing acne. Similarly, other chemical compounds like dioxin may result in the development of acne. Psychological factors such as stress and depression also increase sebaceous gland activity, resulting in acne.

7.2 Assessment of Acne

7.2.1 Acne Vulgaris Grading

There are various grading methods used for the assessment of acne severity. Lehmann et al. [7] reported at least twenty-five such grading methods with different scales. Different grading systems are used in different countries around the world. Every grading method emphasizes each acne type and the different skin regions differently. Using such varying scale systems can assess the severity of acne differently for the same patient. The large number of grading systems available shows that no system is accepted as a standard around the world. The main characteristics of a grading system should include simplicity, accuracy, and efficient assessment.

In the lesion counting method, the number of both noninflammatory lesions, such as comedones, and inflammatory lesions, such as papules, pustules, nodules, and cysts, is recorded. In the photographic method, acne severity is compared against a standard photograph. Carmen Thomas of Philadelphia began using a scoring system for determining the severity of acne vulgaris in the early 1930s, when she started lesion counting in her office notes [8]. In 1956 Pillsbury et al. [9] developed what is considered to be the first grading system. The severity evaluated according to this grading system is based on an overall approximation of the lesion type and number. Table 7.1 explains the grading system developed by Pillsbury et al.

According to this grading system, a patient is graded from 1 to 4 based on the number and type of acne lesions present. For example, an acne patient having comedones and a small number of cysts on the face only is given a grade of 1. Other grades are assigned as shown in Table 7.1. In all grades except grade 4, the acne lesion presence is confined to the face. A grade 4 acne patient suffers from inflammatory and noninflammatory lesions not only on face, but also on the upper trunk.

TABLE 7.1

First Grading Systems [9]

Grade	Description
1	Only the face contains comedones and a rare cyst.
2	Only the face contains comedones, and rarely pustules and cysts that are small in size and number.
3	Only the face contains a large number of comedones and small and large inflammatory lesions.
4	Both inflammatory and noninflammatory lesions appear in large numbers on the face and upper parts of the trunk.

TABLE 7.2

Grading System by James & Tisserand [10]

Grade	Description
1	Only comedones and a small number of papules are present.
2	Comedones, pustules, and a few pustules are present.
3	Larger inflammatory papules, pustules, and a few cysts are present. A more severe form involves the face, neck, and upper portions of the trunk.
4	More severe, with cysts becoming confluent.

In 1958 James and Tisserand suggested another grading method [10]. Their method is presented in Table 7.2. Grade 1 is assigned to a patient if some noninflammatory lesions like comedones and a few inflammatory lesions (i.e., papules) are found on the skin. Similarly grade 2 is assigned to a patient having only three kinds of acne (comedone, papule, and pustule) on the face. Grade 3 and grade 4 are assigned as the count of inflammatory lesions increases and the trunk is included.

In 1966 Witkowski and Simons began using the counting method to assess acne [10]. In this method, each type of acne is recorded and counted. Papules and pustules are differentiated as small and large lesions. Only the right face is included in lesion counting, for time-saving purposes. The acne count for the left side of the face is considered to be same as that on the right side. Nodules and cysts are called "abscesses" in this method. Acne flow and a questionnaire were later included [10].

In 1971 Frank developed a numerical grading of comedones, papules, pustules, nodules and cysts on the face, chest, and back [10]. He suggested two types of grading, from 0 to 4 or from 0 to 10, and a table for record keeping. However, grading was the same as proposed by James and Tisserand (Table 7.2).

In 1975 Plewig and Kligman presented a numerical grading technique [9]. According to this method, comedonal and papulopustular acne are counted and graded separately and overall severity is graded from 1 to 4 based on the number of lesions. The terms comedonal and papulopustular refer to noninflammatory and inflammatory lesions, respectively. Table 7.3 explains, while considering only one side of the face, how the count and type of lesion are related to acne severity.

In 1977 Christiansen et al. presented a 6-point scale based on the reduction percentage of acne lesions. This includes calculating the overall number of lesions and also recording the count of each type of lesion, such as comedones, papules, pustules, nodules, and cysts [10]. The lesion area was used as an assessment area and lesions having a size larger than 5 cm were counted.

TABLE 7.3

Grading Based on Per Half Face Lesion Counting

Grade	Comedonal	Papulopustular
1	<10 comedones	<10 inflammatory lesions
2	10–25 comedones	10–20 inflammatory lesions
3	25–50 comedones	20–30 inflammatory lesions
4	>50 comedones	>30 inflammatory lesions

For each visit of the patient, the change in severity level was evaluated using the 6-point scale, which is presented in Table 7.4.

In 1977 Michaelson et al. [8] counted the number of each type of lesion (i.e., inflammatory and noninflammatory) on the face, chest, and back and assigned a severity score to each type of lesion as shown in Table 7.5.

In 1979 Cook et al. [11] developed an overall numerical grading based on both the photographic standard and 5-point reference scale from 0 to 8. Photographs of each side of the acne patient's face were taken using only one exposure with a special mirror. When the study was completed, the examiners were asked to determine the severity independently. The authors also recommended lesion counting at the initial examination.

Table 7.6 summarizes how a grade (0–8) is assigned to a patient based on the number and type of acne lesions. Grade 0 involves only comedones and papules. The grade (i.e., the severity measure) increases for more severe types of lesions and with a large number of lesions. Grade 8 is assigned to a patient whose face is covered with acne lesions.

Wilson extended the method presented by Cook et al. [10]. He consulted with a large number of dermatologists in a meeting held at the American Academy of Dermatology. In the meeting, these experts evaluated his method.

Burke et al. developed a new 0- to 10-point scale called the Leeds method. The Leeds method comprises a two-way grading system. The first scoring method is applied in clinical practices. It is intended to determine the overall

TABLE 7.4

Scale of Acne Severity Based on Percentage Reduction

Scale	Percentage Reduction	Level
4	100%	Excellent
3	75%–99%	Good
2	50%–74%	Moderate
1	1%–49%	Insufficient
0		Unchanged
−1		Worse

TABLE 7.5

Severity Score and Corresponding Lesion Type

Severity	Type of acne
0.5	Comedone
1.0	Papule
2.0	Pustule
3.0	Infiltrate
4.0	Cyst

severity of a patient's acne. The second scoring method consists of lesion counting. It is intended to be applied during therapeutic examination [10].

Samuelson evaluated acne severity using a photographic method. Nine photographs are used as a reference and acne response to treatment is evaluated in two stages. In the first stage, patients compare their photographs with the nine standard ones. In second stage, experts evaluate the patients using the nine reference photographs. It is mainly concerned with a qualitative description of the relative change of severity during treatment [10]. The relative change and grading are described in Table 7.7.

Lucchina et al. [12] devised another method to determine the severity of noninflammatory lesions like comedones using fluorescent photographs in severity grading. They observed acne on the face only and used a grading scale of 1 to 4 as indicated in Table 7.8 [10].

In 1996 Lucky et al. [13] counted acne lesions on the right and left cheeks, right and left sides of the forehead, and chin; they did not consider the nose. The lesion count was recorded for each facial area. Figure 7.3 is the facial template with the five regions to be observed. Each of the five regions specified in the facial template is observed and the count of each acne type is recorded.

TABLE 7.6

The Cook Grading mMethod [11]

Grade	Description
0	Small scattered comedones and/or small papules.
2	Very few pustules or 3 dozen papules and/or comedones; lesions are hardly visible from 2.5 m.
4	Red lesions and inflammation to a significant degree; worthy of treatment.
6	Loaded with comedones and numerous pustules; lesions are easily recognized at 2.5 m.
8	Conglobata-, sinus-, or cystic-type acne covering most of the face.

TABLE 7.7

Degree of Change and Its Description

Degree of Change	Description
Excellent	There was a decrease of three or more grade numbers with reduced redness and tenderness.
Good	There was a decrease of two grades with reduced redness and tenderness.
Moderate	There was a decrease of one grade with reduced redness and tenderness.
None	There was no change.
Worse	There was an increase of one grade or more or an increase in redness or tenderness with the same grade.

TABLE 7.8

Grading Scale by Lucchina et al. [12]

Grade	Severity
0	None
1	Mild
2	Moderate
3	Extensive

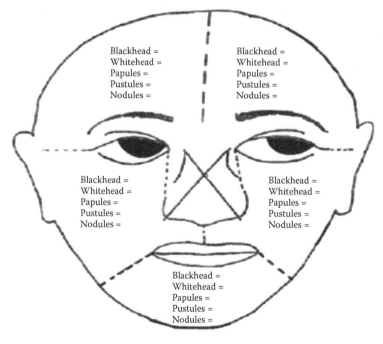

FIGURE 7.3
Facial template [13].

TABLE 7.9

The Global Acne Grading System [14]

Location	Factor (F)	Severity (S)		Local score (F × S)	Acne severity	
Forehead	2	0	Nil		Mild	1–18
Right cheek	2	1	Comedone		Moderate	19–30
Left check	2	2	Papule		Severe	31–38
Nose	1	3	Pustule		Very severe	>39
Chin	1	4	Nodule			
Chest and upper back	3					
		Total Score				

Doshi et al. [14] developed a new method called the global acne grading system (GAGS). In this system, six regions of the human body are observed. The chin, right cheek, left cheek, forehead, nose, chest and upper back are assigned different scoring factors. The regions with a larger area are assigned a higher factor. Similarly, each acne type is assigned a different weight. For the assessment of overall severity, the face region factor is multiplied by the severity factor. The product of these two factors results in a local score for the region. A total score is then calculated by summing all the local scores. The patient's acne severity is assessed by comparing the total score against the acne severity standard. The acne severity is mild if the total score is 1–18, moderate is 19–30, severe is 31–38, and very severe is 39 or greater. Table 7.9 explains the GAGS.

The six regions, numbered I–IV in Figure 7.4, are observed for the assessment of acne. As shown in Table 7.9, each of these regions has a local factor. This measure indicates the importance of the area and its surface area as well.

O'Brien et al. reviewed the Leads acne grading system. In their revised system, the face, chest, and back are used for severity assessment [15], as well as patient comparisons with the standard photographs. As shown in Table 7.10, there are three ranges of grades and corresponding severity for the face and chest and back. In developing this revised system, three dermatologists and four acne assessors worked on 1,000 photographs.

Hayashi et al. devised a four-grade system. Their method includes the use of both reference photographs and lesion counting. Based on these two measures, the acne patient can be classified into one of the four grades [16]. As shown in Table 7.11, acne severity is considered mild for a total lesion count of 0 to 5. The severity is considered moderate for a lesion count of 6 to 20, severe for a count of 21 to 50, and very severe for a count of more than 50.

Eichenfield et al. [17] modified the GAGS described by Doshi et al. [14]. Using this modified grading system, they investigated the efficacy of the tretinoin pump for facial acne therapy at stage 4. As in the GAGS, the total score in the modified GAGS is calculated by summing the products of regional

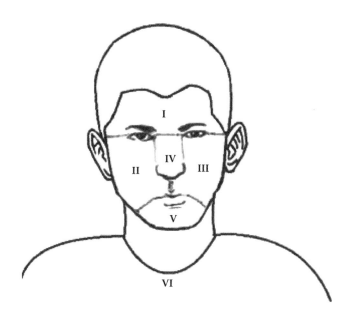

FIGURE 7.4
The six regions (I–IV) used in the GAGS.

TABLE 7.10

Grading of Acne Severity Based on a Photographic Standard [15].

Body part	Mild	Moderate	Severe
Face	Grade 1 to grade 4	Grade 5 to grade 8	Grade 9 to grade 12
Upper chest and back	Grade 1 to grade 3	Grade 4 to grade 5	Grade 6 to grade 8

TABLE 7.11

Grading Based on Lesion Counting and Reference Photographs

Lesion Count	Severity Grade
0–5	Mild
6–0	Moderate
21–0	Severe
>50	Very severe

TABLE 7.12

Modified Global Acne Grading System [17]

Location	Factor (F)	Severity (S)	Local score (F × S)	Acne severity	
Forehead	2	0	Nil	Mild	1–18
Right cheek	2	1	Comedone	Moderate	19–30
Left check	2	2	Papule	Severe	31–38
Nose	1	3	Pustule	Very severe	>39
Chin	1	4	Nodule		
		Total score			

factors and severity factors. Acne severity is evaluated based on the total score obtained for each patient. The ranges and the corresponding severity are shown in Table 7.12.

As mentioned earlier in this section, different grading systems are used for the assessment of acne severity in different countries around the world. In these grading systems, different severity scores are assigned to each acne type and each region of the human skin. The grading systems along with the countries in which they are used are summarized in Table 7.13.

7.2.2 Technologies for the Assessment of Acne Lesions

In clinical practice, acne severity is evaluated either by direct visual inspection of the acne patient or by comparison with a standard photograph. The use of fluorescence photography was introduced for the first time in 1996 for evaluating acne severity. About forty subjects with different grades of acne vulgaris were chosen and their photographs were taken using flash and fluorescence at baseline and at intervals of four, eight, and twelve weeks [12].

TABLE 7.13

Grading Systems and Corresponding Countries of Use

Country	Grading system
Hong Kong	Global Acne Grading System [18]
India	Global Acne Grading System [19]
Japan	Hayashi et al. [16]
Jordan	Global Acne Grading System [20]
Korea	Korean Acne Grading System [21]
Malaysia	mGAGS and Leeds Grading System [22]
Saudi Arabia	Global Acne Grading System [23]
Turkey	Global Acne Grading System [24]
United Kingdom	Leeds Grading System [25]
United States of America	Investigator's Global Assessment (IGA) [26]

(a) Flash
photograph

(b) Fluorescence
photographs

FIGURE 7.5
Lesion detection using photography: (a) flash photograph; (b) fluorescence photographs.

Different equipment (e.g., Minolta X-700 camera, Vivitar series I 28–105 mm macro lens, and Tamron 90 mm macro lens) was used to take simple flash photographs. In fluorescence photography, the same camera was used, but in order to reduce the effects of ultraviolet light, a special filter was used with the camera. The Kodak Wratten #4 filter was found to have the desired properties. In Figure 7.5, both flash and fluorescent photographs are shown. In Figure 7.5(b), the dark regions represent the different kinds of acne lesions.

For the assessment of both inflammatory and noninflammatory lesions, Phillips et al. studied the use of polarized light photography [27]. They attempted to determine the degree of agreement between the acne assessment obtained from clinical evaluation and the assessment obtained from photographs captured with flash photography and perpendicular polarized light photography. For flash photography, a Minolta X-700 camera was used with a Tamron macro lens. In polarized photography, a Minolta 80PX was used with a ring flash and linear polarizer.

In order to keep the camera and flash position the same, a photographic table was used and the photographs were taken from the front. During image acquisition, the subjects were directed to look into the camera. The right and left cheeks were imaged at an angle of 45° using the special polarizer.

Both flash and perpendicular polarized light photographs are shown in Figure 7.6. Different skin features, such as color, are enhanced visually. With this setup and processing, the smallest acne lesions (i.e., comedones) are visible in the flash photograph and thus are easy to identify and count. In perpendicular polarized light photographs, the comedones are significantly different from the skin tone.

In 2001 Rizova and Kligman made use of crossed and parallel polarizing light photography along with sebum production measurements. In this research the effect of adapalene gel 0.1% on all kinds of acne vulgaris lesions

(a) Flash photograph   (b) Perpendicular polarized
                                 light photograph

FIGURE 7.6
Comparison of flash and perpendicular photography: (a) flash photograph; (b) perpendicular polarized light photograph.

was examined. The sebaceous glands showed a significant response to the adapalene gel 0.1% with lower sebum production. Thus adapalene gel is used for all kinds of acne lesions and is also very effective in preventing the development of new lesions [28].

In 2008 Do et al. examined computer-assisted alignment and tracking of acne lesions [29]. Images were taken using a Nikon D1x digital camera at a resolution of 1960 × 3008 pixels (5.9 megapixels) and a fixed reproduction ratio equivalent to 1:6 on 35-mm film. The digital photographs were modified with editing software for better alignment and tracking. The photographs of the selected acne patients were taken at intervals of two weeks for a total of twelve weeks. Skin regions with dense and large numbers of lesions were taken as regions of interest (ROIs).

Such ROIs were examined in twenty-seven acne patients with mild to moderate acne in accordance with the Leeds grading system. A set of seven ROIs was recorded for each patient. The number of mild to moderate acne lesions was determined. Next, the lesions were classified into different types (i.e., noninflammatory [comedones] and inflammatory [papules, pustules, and nodules]). The size of each ROI is 400 × 400 pixels and all seven were printed on a single A4 (paper size) template.

The experts assigned a numerical value to all acne after tracking the inflammation of the lesion (image 0 in Figure 7.7). The acne lesion growth was observed at a regular interval of two weeks. This technique eradicated the conflicts found in standard photographs, such as changes in camera angle and the need for registering two frames. Figure 7.7 illustrates the procedure of labeling and tracking acne changes.

Multispectral imaging (MSI) techniques are also used for the assessment and evaluation of acne lesions. In MSI, the images of human skin are captured at wavelengths lying outside the visible spectrum. With MSI, one can

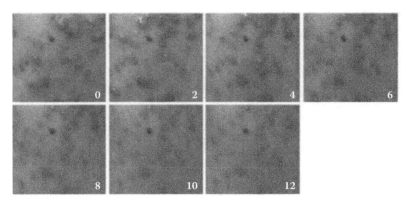

FIGURE 7.7
Tracking of acne changes.

capture an image with more information than a simple color image. MSI provides data at specific wavelengths across the electromagnetic spectrum. Also, multispectral images do not have color and illumination problems. MSI is widely used in the biomedical field [38].

Spectral information such as the power distribution of a reflected wavelength, instead of color information, was used for the identification and classification of acne lesions. In 2008 the pioneer of this concept, Hideaki Fujii, used the images captured with a 16-band multispectral camera for spectral feature extraction [30]. The complete image acquisition setup consisted of a 16-band multispectral camera (16 bands, 12 bits, 2048 × 2048 pixels) and two tungsten lamps as a light source.

The artifacts introduced in images during the acquisition process are removed in a preprocessing step and spectral information is calculated for each pixel for classification purposes. The relative reflectance obtained for an object is a function of the 3D appearance of the object and suffers from artifacts such as shades and gloss. In order to remove the shades and shadows, the resulting multispectral signal of the image is normalized by division with total intensity. Different mathematical models are used for the identification and classification of each type of acne. The mathematical models used are called linear discriminant functions (LDFs). In order to classify the acne lesions among the three classes, three Fisher LDFs are used with three threshold values determined through experimentation. With the help of MSI and LDF classifiers, they succeeded in characterizing and classifying several skin lesions, such as comedones, reddish papules, pustules, and scars.

In order to analyze the skin lesion objectively, Bae et al. [31] developed a multimodal imaging modality with which color images of different kinds can be captured. With multimodal facial color imaging, three kinds of photographs (simple color images, parallel, and cross-polarized) can be taken. The main use of fluorescent images is in the quantitative evaluation of sebaceous glands.

7.2.3 Objective Measurement of Acne Lesions

From the discussions in the previous sections, it can be seen that there are various grading methods for acne vulgaris. All the grading methods assess acne lesions, hence methods for the objective measurement of acne lesions will allow accurate acne grading. However, as discussed in section 7.3, no such method currently exists.

In clinical practice, two methods are used for the assessment of acne lesions: lesion counting and comparison with standard photographs. Dermatologists determine the severity of acne using a specific grading system and suggest proper treatment, keeping in the mind the severity level of the acne. However, this assessment is subjective and involves both interrater and intrarater variability. Intrarater variability means that the same patient may be graded differently by the same dermatologist at different times. Interrater variability means that a patient may be graded differently by different individuals. In addition, lesion counting is a tedious and time-consuming task. Thus a system to objectively measure acne lesions is very important to provide an accurate acne grade and improve treatment efficacy.

Acne has diverse impacts on patients, including feelings of ugliness, low self-esteem, poor social life, psychological effects, and itchiness. To minimize these effects, acne must be treated properly and at the proper time. However, proper treatment is highly correlated with an accurate assessment and grading of the acne severity level.

To overcome these problems, an acne assessment system is needed that can assess acne severity utilizing imaging processing techniques. This system involves five steps: image acquisition, enhancement, segmentation, feature extraction and selection, and classification and grading. In the image acquisition step, images of the patient are taken under proper lighting conditions using a digital camera. In the image enhancement process, two functions are performed. First, artifacts such as nonuniform illumination and blurring due to motion of the subject and camera, introduced during image acquisition, are removed. Second, the image is processed in order to enhance the features of interest in the image.

In segmentation, the objects of interest (lesions) are separated from the background (skin). In the feature extraction and selection process, all those features having good discriminating power are extracted. In the classification step, the lesions are classified into different types by a trained individual. Once the lesion count is available, the severity can be evaluated using a specific grading system, such as the mGAGS. This system can remove the interrater and intrarater subjectivity, producing an accurate assessment in less time.

7.3 Automatic Assessment of Acne

7.3.1 Image Acquisition Setup

Image acquisition is the first and most important step in the development of a computer-based automated system. In order to take high-quality, artifact-free images, it is necessary to know the different aspects of image acquisition (e.g., what kind of camera should be used, proper lighting conditions, and how the distribution of light sources and intensity of light effect the appearance of objects in images). Cameras with different parameters (e.g., resolution, auto white balance, ISO speed, and aperture size) are capable of capturing images with different qualities. There is a strong relationship between the values of the camera internal parameters and the quality of the images. This application demands high-quality detailed images that can be captured only with a high-resolution camera. Most digital single-lens reflex (DSLR) cameras meet these criteria. Camera resolution refers to the number of pixels (picture elements) used to represent a digital image.

For the development of an automatic acne assessment system, detailed images of high quality are needed. In order to take such an image, a Canon 500D was used. It is a 4752 × 3168 pixel camera with charge-coupled device (CCD) sensors. Close-up images were taken with a constant distance of three feet between the patient and the camera.

The quality of the images is also affected by the lighting conditions under which the images are taken. Lights of different intensities have different color temperatures and produce different color shades on image. Some cameras have the ability to estimate the color temperature of the available light source and correct the color of the image. A nonuniform distribution of light sources produces artifacts like shadows, specular illumination, and nonuniform intensity distributions. Similarly, the background and its roughness can also affect image quality.

To capture good quality images, the light sources must be properly distributed to obtain a diffuse, uniform light on the object. Soft-boxes can be used in front of light sources in order to get diffused light directed at the subject. The light sources and soft-boxes used in the setup are shown in Figure 7.8. Shadow effects can be minimized by using reflectors held at a proper distance from the patient. A neutral color or green background is often used in digital photography.

7.3.2 General Methodology

As mentioned, there are more than twenty-five acne grading systems. However, the GAGS and the mGAGS (most widely used by dermatologists in Malaysia) were used in the development of this automatic assessment system. In 1997 Doshi et al. [14] introduced the GAGS. Six different regions are

FIGURE 7.8
Lighting and background setup.

assessed separately in this grading system (forehead, chin, left cheek, right cheek, nose, upper chest and back). Each region has a dedicated score (F), as shown in Table 7.9. The lesion types are also assigned scores according to their severity (S). Each of the regions is graded separately by determining the most severe lesion type present in that particular region. The local score is then calculated by multiplying the region factor (F) by the severity score (S) of the most severe lesion in that region. All local scores are then summed to find the total score. For the overall severity, the total score is compared with the ranges of the acne severity criteria shown in Table 7.12. In 2008 the GAGS was modified (mGAGS) by Eichenfield et al. by removing the nonfacial regions from the assessment criteria (i.e., chest and back) (Table 7.13).

The automated acne vulgaris grading system has the six system components shown in Figure 7.9. The first step is image acquisition. Object detection becomes easier and more efficient if the image has a high resolution, details are sharp, and lighting is proper and uniform. Thus, for better accuracy in the assessment, image acquisition plays a crucial role. After acquiring an image the contrast of the image is improved for enhancement of details in the image. This step is characterized as a preprocessing step. Preprocessing is not the detection procedure, but it is crucial in making further processing (like object detection and recognition) easier and more efficient. In automated diagnosis systems for lesions, the preprocessing step is generally the step prior to detection/segmentation of a lesion. Thus preprocessing is generally done by transforming the image into a suitable color model, by enhancement of contrast, by removal of artifacts, or by noise reduction. The segmentation algorithm is applied next. In the segmentation step, the image is partitioned

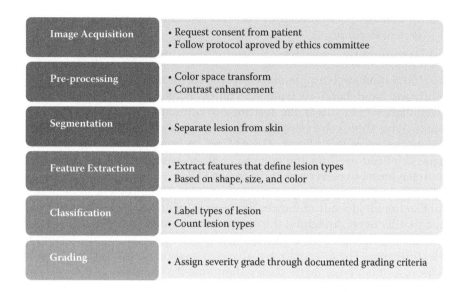

Image Acquisition	• Request consent from patient • Follow protocol aproved by ethics committee
Pre-processing	• Color space transform • Contrast enhancement
Segmentation	• Separate lesion from skin
Feature Extraction	• Extract features that define lesion types • Based on shape, size, and color
Classification	• Label types of lesion • Count lesion types
Grading	• Assign severity grade through documented grading criteria

FIGURE 7.9
Proposed framework for an automated grading system of acne lesions.

into various segments in order to extract meaningful information from the image. The pixels that have common characteristics with respect to color, intensity, or texture are assigned identical labels. From the segmented regions of the image, features based on size, shape, color, or texture are extracted for each region. Based on those features, an object is classified and recognized. Once the lesion objects are recognized, grading of the acne vulgaris lesion is computed through any of the acne vulgaris grading criteria.

7.3.3 Enhancement Methods

The contrast enhancement discussed in this chapter is compared with other methods such as decorrelation contrast stretching, histogram equalization (histEq), and contrast enhancement through local rank transformation (LRT) [35]. Decorrelation stretching is a technique that involves the mapping of original color values of the image to a new set of color values with a wider range. The color intensities of each pixel are transformed into the color eigenspace of the NBANDS × NBANDS covariance or correlation matrix, stretched to equalize the band variances, then transformed back to the original color bands [36]. There are four steps to perform decorrelation, summarized as follows:

1. Extract the principal components of the image.
2. Rotate and translate along the axes of the principal components.

3. In this new domain, apply contrast stretching for dynamic range expansion of pixels.

4. Transform the image back to original coordinates for display.

Decorrelation stretching enhances the color separation of an image with significant band–band correlation. The overstated colors improve visual interpretation that can make feature discrimination easier. Color images usually have three color bands, but decorrelation can be applied regardless of the number of color bands (NBANDS). In image processing, this technique has been used for the enhancement and stretch of color differences (contrast) found in each pixel of the image [37].

A very commonly used method for contrast enhancement is histogram equalization. It globally enhances the contrast of an image having close contrast by effectively stretching the most repeated intensity values. Histogram equalization increases the contrast of the image by stretching the histogram to its full range. The histogram equalization method was chosen for the comparison, as it is a basic method for enhancement of images.

LRT was chosen as the algorithm for enhancing the results achieved by this method. Mukherjee [35] discussed the interesting properties of rank transformation. In this work, pixel intensities were transformed to their local rank for the purpose of edge extraction and enhancement, thus it is referred to LRT. The precise mathematical definitions are discussed as well. This section only provides an overview and summary of his work.

The local rank of an element e_i in a set U is defined as the number of elements of Si that are less than e_i:

$$U = \{e_1, e_2, e_3, e_4, e_5\}.$$

Suppose $e_1 < e_4 < e_2 < e_3 < e_5$, then

$S1 = \{e_1, e_2\}$	implies local rank $(e_1) = 0$
$S2 = \{e_1, e_2, e_3\}$	implies local rank $(e_2) = 1$
$S3 = \{e_2, e_3, e_4\}$	implies local rank $(e_3) = 2$
$S3 = \{e_3, e_4, e_5\}$	implies local rank $(e_4) = 0$
$S4 = \{e_4, e_5\}$	implies local rank $(e_5) = 1$

LRT$(U) = \{0, 1, 2, 0, 1\}$, where LRT(U) denotes the LRT of set U.

In this example $n = 5$ and $k = 1$. The δ-rank transform of an element e_i in set U is the number of elements in set U that are less than element e by at least δ. It has been observed that a positive δ in LRT extracts the edges, whereas with a negative δ smooth areas are highlighted.

The image is first transformed to the YCbCr color space and LRT is applied to the Y component. Equation (7.1), as described in Mukherjee [35], is used to compute the enhanced image I:

$$I = Y + \lambda * LRT(Y, \delta1) + \gamma * LRT(Y, \delta2), \text{ where } \delta1 < 0 \text{ and } \delta2 > 0. \quad (7.1)$$

λ and γ are used to provide weight to the smooth details and edge details, respectively. These details are added to the original luminance component Y, resulting in an enhanced image I.

7.3.4 Metrics Used for Quantitative Analysis

The objective of the comparative study was to measure the quality of the enhanced image produced by our proposed methodology compared with that of selected methods. Two objective measures were set up to compare the results of the image enhancement process: the contrast improvement factor (CIF) and image contrast normalization (ICN).

Through contrast enhancement, it was intended to increase the contrast between the lesion and nonlesion areas. Thus the contrast between acne lesion and the normal skin could be defined as the absolute mean intensity difference between the lesion pixels and nonlesion pixels in an image. Mathematically, the contrast between lesion and nonlesion areas can be defined by equation (7.1):

$$C_{|L-nL|} = \left| \frac{1}{p} \sum_{i=1}^{p} IL_i - \frac{1}{q} \sum_{j=1}^{q} InL_j \right| \qquad (7.1)$$

IL and InL are the pixel intensity values of the lesion and nonlesion areas, respectively. The values of p and q are the number of pixels representing the lesion and nonlesion areas, respectively. The higher the value of $C_{|L-nL|}$, the better the contrast of the lesion in the image will be.

The CIF is the ratio of the contrast of the object (lesion) and the background (nonlesion) obtained by a specified algorithm (C_{alg}) and that of the reference (C_{ref}), as follows:

$$CIF = \frac{C_{alg}}{C_{ref}} \qquad (7.2)$$

Here, C_{ref} is the contrast between the lesion and nonlesion areas in the green band of the original image. The green band was chosen as the reference because the green band provides a good contrast between lesion and nonlesion compared with other bands in a color image.

The standard deviation (σ) shows the spread of the data from the mean value. A low standard deviation specifies that data are mostly concentrated toward the mean value. Thus, in terms of image intensity, a low standard deviation implies that most of the intensity values are close to the mean and are not spread out over a wide range. Hence the homogeneity of the intensity values can be indicated by standard deviation as

$$\sigma = \frac{1}{mn} \sum_{i=1}^{(mn)} (I_i - \bar{I})$$

$$(7.3)$$

The general aim when measuring contrast is that the variation of intensity values of the overall image should be low while the difference in object intensity values from that of the background is high. Therefore ICN can be formulated as follows:

$$ICN = \frac{\frac{1}{mn} \sum_{i=1}^{(mn)} (I_i - \bar{I})}{C_{|Les-nLes|}}$$

(7.4)

where I_i and $\bar{I}$ represent the intensity of the ith pixel and mean intensity of the image, respectively.

Having incorporated both standard deviation and average contrast, ICN not only measures the contrast normalization, but also the contrast enhancement. The lower the ICN is, the better the contrast normalization of the image will be. This criterion can also be used to measure noise reduction due to a specified image enhancement process. Noise causes variations in the intensity of an object. Ideally an image without any noise would have a uniform intensity distribution. If the ICN value obtained is small enough, meaning the standard deviation is also small, the more homogeneous the intensity distribution will be.

7.3.5 Enhancement Method for Acne Grading

The contrast enhancement process is carried out by increasing the dynamic range of the pixels. Dynamic range is defined as the ratio between the smallest and the largest possible values of a changeable quantity. For an imaging device or camera, it is measured in bits. Thus for an 8-bit monochrome camera, there are only $2^8 = 256$ different levels at which intensities of a scene can be represented. But if there are extra details present, a display device will forcefully map them to these 256 levels, as a result of which either those details will be completely lost or their difference from the surrounding details will be reduced.

A common way to produce a high dynamic range (HDR) image of a scene is to capture the low dynamic range (LDR) at multiple exposure settings. At low exposure, details in the bright region are visible, while at high exposure the details in the dark region are visible. In the proposed method, LDR images are generated artificially using LRT. The enhancement method creates LDR images artificially, that is, instead of producing multiple LDR images by changing the exposure settings of the camera, artificial images are created using LRT. Hence different details of the LDR image are highlighted. The HDR image includes the details of all the LDR images from which it is approximated. Logarithmic tone mapping is then applied to make it displayable on an LDR device. Figure 7.10 depicts the steps involved in our contrast enhancement methodology.

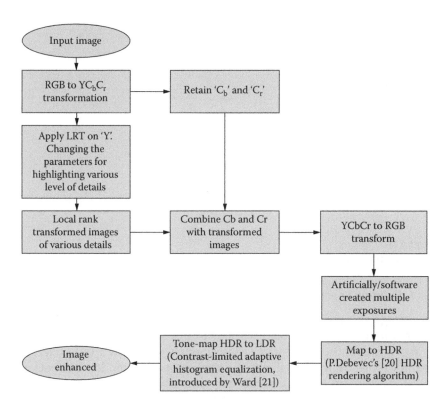

FIGURE 7.10
Process flow diagram of the enhancement method.

7.3.6 Segmentation method for Acne Grading

A novel and straightforward approach for the segmentation of acne vulgaris lesions has been developed. The algorithm tends to classify the pixels purely based on the spectral range of its color. In other words, no local or global information is used while performing the segmentation. An effective technique was proposed by Smits [38] for a red, green, blue (RGB) to spectrum conversion of reflectance. The linearity of the conversions is exploited to create a spectra of primary colors (red, green, and blue), secondary colors (cyan, magenta, and yellow) and white. Any color can be expressed as a sum of white plus one of the primary colors plus one of the secondary colors.

Visual analysis of acne lesion images reveals that pixels should be classified into categories of red lesion, purple lesion, brown lesion, scar, and specular reflection. The pixels that belong to the first three categories are classified as lesion pixels, while others are classified as nonlesion pixels. The segmentation algorithm proceeds as follows:

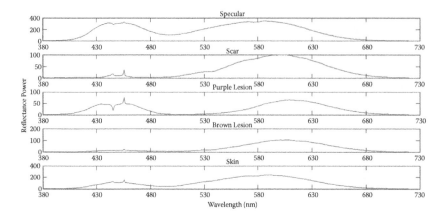

FIGURE 7.11
Average spectral models.

Step 1. Building Spectral Models for Color Categories

The average spectral model is calculated from the sample spectral models that are manually built through various example lesion images for each of the above mentioned categories. These models are created only once and saved for reuse. Figure 7.11 shows the average spectral models built of various categories. Spectral models are one-dimensional curves of reflectance power of the wavelengths from 380 nm to 720 nm with steps of 1 nm.

Step 2. RGB to Spectral Reflectance Conversion

An RGB input image of the affected area of a lesion is first converted into the spectral domain (380–720 nm with steps of 1 nm) [38]. In this way the spectral curve for each pixel in the image is obtained.

Step 3. Mahalanobis Distance Classification

The spectral curve for each pixel is a test point that has to be classified into one of the classes (i.e., red lesion, purple lesion, brown lesion, scar, or specular reflection). Then, for each of the given test points, the Mahalanobis distance is computed for each class, classifying the test point as belonging to that class for which the Mahalanobis distance is minimal. The Mahalanobis distance is a statistical metric for gauging the similarity of an unknown pattern with a known pattern. It takes into account the correlation between two variables and is scale invariant. To find the dissimilarity between two random variables a and b, the Mahalanobis distance is mathematically expressed as equation (7.5):

$$d(a,b) = \sqrt{\frac{(a-b)^T (a-b)}{S}} \tag{7.5}$$

Step 4. Binary Classification of Lesion and Nonlesion

Finally, a binary classified/segmented image is obtained in which pixels belonging to the first three classes (i.e., red lesion, purple lesion, and brown lesion) are classified as lesion pixels and the rest as nonlesion pixels.

7.4 Results and Discussion

The quantitative or statistical results are provided in Tables 7.14 and 7.15.

For ease of interpretation and analysis, Figures 7.12–7.14 show the graphical representation of numerals in Tables 7.14 and 7.15. In Figure 7.12, the vertical axis represents the mean CIF value, whereas the horizontal axis refers to the five regions of the face (i.e., left cheek [LC], right cheek [RC], chin [C], nose [N], and forehead [F]). The legend shows each of the five contrast enhancement approaches.

Our proposed method displays a better CIF value than the simple LRT method. Among all the techniques, decorrelation and histogram equalization (histEq) are quite competitive in showing good CIF values. It can also be observed through qualitative results that both these methods generate high contrast results.

The nose region has the lowest CIF values in all methods and the cheeks have higher CIF values as compared with other regions. The reason is that lesions on the nose are usually not as dark as on the cheeks.

ICN is the ratio of the standard deviation of the intensity value to the difference in mean intensity values between the object and the background. Therefore, a lower value of ICN is preferable. Figure 7.13 shows ICN values for all five regions collectively. The vertical axis represents the mean ICN value, whereas the horizontal axis refers to the five regions of the face (i.e., left cheek [LC], right cheek [RC], chin [C)] nose [N], and forehead [F]).

TABLE 7.14

Comparison of Constrast Improvement Factor (CIF)

Region	Our Method	Decorrelation	HistEq	LRT
Left cheek	4.16	4.35	6.48	3.53
Right cheek	3.86	7.80	7.36	2.00
Chin	1.72	3.13	5.37	1.31
Nose	0.89	1.55	3.85	0.81
Forehead	3.04	3.24	5.04	2.15
All regions	2.74	4.01	5.62	1.96

TABLE 7.15

Comparison of Image Contrast Normalization (ICN)

Region	Our Method	Decorrelation	HistEq	LRT
Left cheek	1.61	1.50	2.72	1.77
Right cheek	2.02	3.37	5.53	3.16
Chin	1.28	1.75	2.69	1.62
Nose	1.29	1.20	2.52	2.10
Forehead	1.75	1.39	2.99	1.46
All regions	1.98	3.15	3.29	2.02

The bar plot in Figure 7.14 shows the ICN values of all five regions collectively. The vertical axis represents the mean ICN value and the horizontal axis refers to the enhancement methodology.

Among the four methodologies, the proposed method keeps a balance of the standard deviation of intensity values and the difference of mean intensity values between lesion and nonlesion areas such that it provides the lowest ICN value. Histogram equalization has the highest ICN value due to the high variation in intensity values. LRT is very close to the proposed method in ICN values, however, it has a very low CIF value.

Next, the segmentation was applied over the enhanced image to differentiate the lesion from the skin. The proposed segmentation method was tested quantitatively for specificity and sensitivity. The segmentation algorithm was not only tested on the proposed enhancement method but on other selected enhancement techniques as well. The quantitative results provided are the average for the entire dataset.

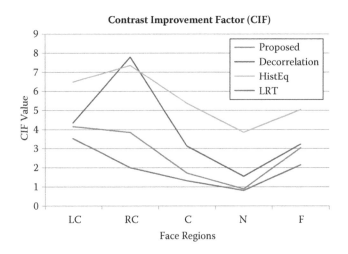

FIGURE 7.12
Regional mean plot of the CIF.

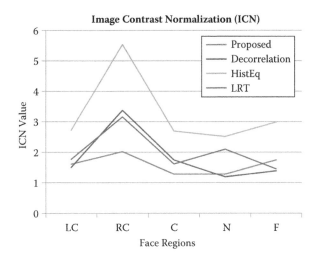

FIGURE 7.13
Regional mean plot of ICN.

Table 7.16 is a quantitative comparison of the segmentation approach applied over the contrast enhancement method and two other selected methods. This was done to ensure the efficacy of the contrast enhancement approach along with segmentation, as the purpose of the contrast enhancement was to achieve effective segmentation results.

Table 7.16 shows the average sensitivity and specificity of the algorithm with respect to the acne severity grade. The specificity comparison was quite satisfactory in almost all cases. The method shows a specificity of greater

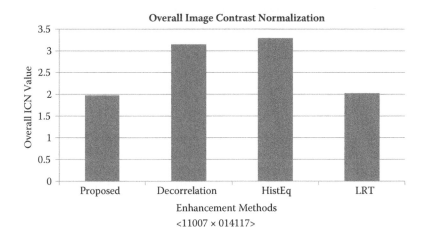

FIGURE 7.14
Overall ICN.

TABLE 7.16

Results of Segmentation

Severity	Method	Sensitivity	Specificity
Mild	Our method	62.65	80.91
	Decorrelation	65.83	79.41
	histEq	61.01	75.54
	LRT	53.47	85.62
Moderate	Our method	66.52	83.17
	Decorrelation	59.00	82.88
	histEq	58.31	75.78
	LRT	44.78	86.78
Severe	Our method	72.03	84.32
	Decorrelation	47.12	82.01
	histEq	48.24	75.65
	LRT	36.74	86.10
Very severe	Our method	70.71	85.50
	Decorrelation	48.75	84.52
	histEq	58.56	76.99
	LRT	39.79	83.68

than 80%. The specificity shown by the histogram equalization approach is greater than 75% and LRT shows a specificity of greater than 80% in all severity cases. The decorrelation method shows a specificity of greater than 79%. This shows that the ability of the method to identify nonlesion pixels is satisfactory as compared with other methods. However, it should be noted that specificity decreases as the severity grade increases. Figure 7.15 illustrates some segmentation results of lesion images along with their respective sensitivity and specificity values.

Sensitivity results show the ability of the algorithm to detect positive lesion pixels. The proposed segmentation method has far better sensitivity compared with the other two methods. The sensitivity score is best with the severe cases of acne and worst with mild cases. The reason that the severe criteria show better results is because acne lesions have good contrast with the skin at that stage. In mild cases, the lesions are hardly visible and lesion types other than comedones are not present. Therefore the sensitivity result is poor in mild cases. In very severe cases, the skin surface is disrupted with cystic and nodular lesions. In most cases there is no defined boundary of such lesions due to the erythema surrounding them, thus the sensitivity is reduced in very severe cases. This analysis shows that the proposed method for contrast enhancement fits best as a preprocessing step to our proposed segmentation approach.

S. No.	Original	Our Method	Decorrelation	HistEq	LRT
1	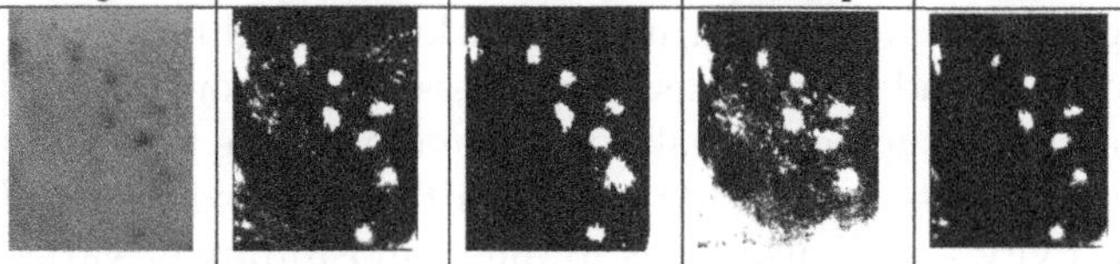				
	Sensitivity Specificity	88% 91.9%	88.99% 94.6%	88.85% 73.22%	76.13% 97.15%
2	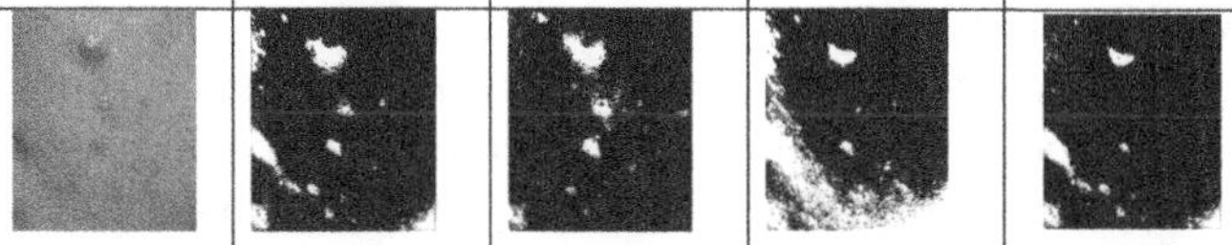				
	Sensitivity Specificity	73% 93.3%	71.25% 91.35%	54.13% 75.56%	44.75% 95.24%
3	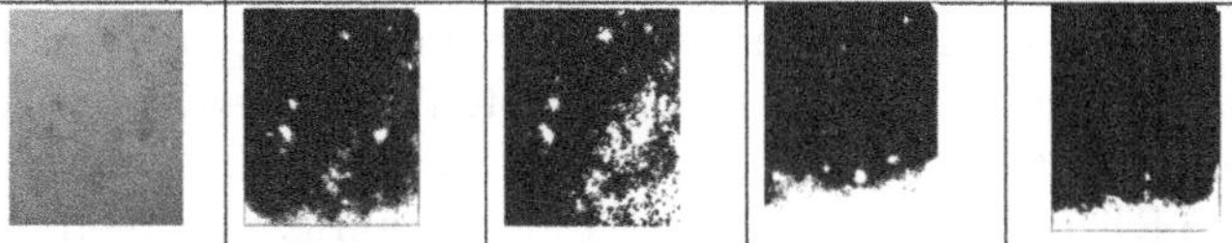				
	Sensitivity Specificity	67.79% 90.2%	74.2% 89.83%	60.08% 76.89%	44.14% 84.69%
4	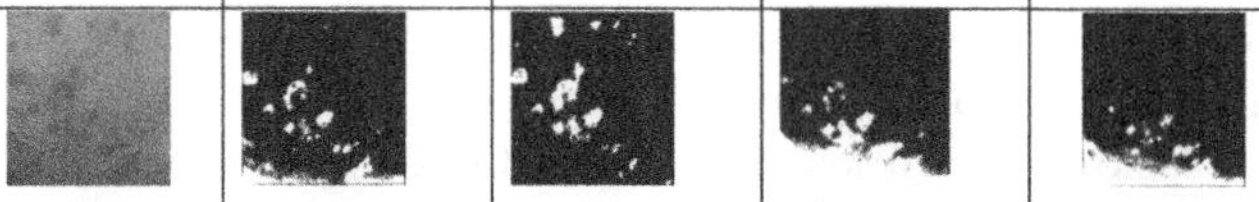				
	Sensitivity Specificity	64.35% 92.1%	58.05% 76.18%	70.5% 77.35%	42.10% 84.98%
5					
	Sensitivity Specificity	86.59% 95.37%	64.36% 72.19%	80.32% 78.24%	68.26% 90.25%

FIGURE 7.15
Qualitative/visual comparison of lession images.

7.5 Conclusion

The work in this chapter concerns the development of a computerized diagnosis system for identifying the severity of acne vulgaris lesions. There are more than twenty-five acne vulgaris grading schemes, as discussed in detail in section 7.2.1. All the grading schemes are based on the assessment of six different types of acne lesions (i.e., identification of the type of acne lesion and/or the count of the particular acne lesion type). This is done for the various parts of face (forehead, nose, chin, left and right cheeks) as well as for the upper back and chest. Thus the process is time consuming and tedious. As a result, dermatologists make an estimation of the type and number of acne lesions and assign a grade based on that approximation. This introduces subjectivity and it also leads to inter- and intrarater variability. Therefore in this chapter we introduced the initial essential steps for an automated system that are required for the elimination of errors caused by human subjectivity in identifying and counting acne lesions.

The automated system for acne grading is based on capturing the images of various body parts using a DSLR camera. The background is kept uniform while diffused lighting is used during image acquisition. The image acquisition setup is described in detail in section 7.3.1. After image acquisition, the acne lesion images are processed for contrast enhancement and segmentation of the lesions. Contrast enhancement and segmentation are two of the key and essential steps of the automated acne grading system. The contrast enhancement is done by artificially generating HDR images of acne lesions using LRT followed by log-based tone mapping. This method was compared with histogram equalization and decorrelation stretch methods. The results showed that our method outperforms the other methods both qualitatively and quantitatively based on the CIF and ICN.

In this chapter we also present the acne lesion segmentation method based on the spectral models of acne lesions. The spectral models include red lesion, purple lesion, brown lesion, scar, and specular reflection. The image pixels defined by the scar and specular reflection models are classified as nonlesion pixels, while those defined by the red, purple, and brown models are classified as acne lesion pixels. The sensitivity and specificity results show that our method performs better than the histogram equalization and decorrelation stretch methods. The next steps for successful implementation of the automated acne grading system are feature extraction and classification, as shown in Figure 7.9. Currently we are in the process of developing these two steps and our future plans include clinical trials of the automated acne grading system.

Appendix: Algorithm MATLAB code

```
%%%%%% main.m %%%%%%
%%%%%%%%%%Build Artifitial Exposures%%%%%%%%%%
I = imread('img.jpg');% input image
disp('Creating Artificial Exposures....');
X = CELRT(I,11,11,1,0,25,-25);imwrite(X,'exp2.jpg');
X = CELRT(I,11,11,0.8,0.2,25,-25);imwrite(X,'exp1.jpg');
```

```
X = CELRT(I,11,11,0.5,0.5,25,-25);imwrite(X,'exp.jpg');
X = CELRT(I,11,11,0.2,0.8,25,-25);imwrite(X,'exp-1.jpg');
X = CELRT(I,11,11,0,1,25,-25);imwrite(X,'exp-2.jpg');
disp('Artificial Exposures created.');

%%%%%%%%%%Apply Tone Mapping%%%%%%%%%%
files = {'exp2.jpg', 'exp1.jpg', 'exp.jpg','exp-1.jpg', 'exp-2.jpg'};
expTimes = [1 1.2 1.44 1.728 2.0736];

hdr = makehdr(files, 'RelativeExposure', expTimes);
rgb = tonemap(hdr);
h = fspecial('gaussian', [25 25], 2.5);
filtRGB = imfilter(rgb, h);
imwrite(filtRGB,'outputImg.jpg');
disp('Enhance Image generated by name "outputImg.jpg"');

%%%%%%%%%%Deleting Exposure image files%%%%%%%%%%
delete('exp2.jpg');
delete('exp1.jpg');
delete('exp.jpg');
delete('exp-1.jpg');
delete('exp-2.jpg');
disp('Exposure image files deleted...');
```

%%%%% **CELRT.m** %%%%%

```
function [out] = CELRT(I,r,c,L,G,z1,z2)

    YCBCR = rgb2ycbcr(I);
    Y = YCBCR(:,:,1);
Cb = YCBCR(:,:,2);
    Cr = YCBCR(:,:,3);

    [LRT1] = LRT(Y,r,c,z1);
    [LRT2] = LRT(Y,r,c,z2);

    X = Y+(L.*LRT1)+(G.*LRT2);

Cbm = uint8((double(X./Y).*double(Cb-128))+128);
Crm = uint8((double(X./Y).*double(Cr-128))+128);

out = ycbcr2rgb(cat(3,X,Cb,Cr));

end
```

%%%%% **LRT.m** %%%%%

```
function [out,norm] = LRT(I,r,c,z)
    Rows = r;
    Cols = c;

    m = size(I,1);
    n = size(I,2);

prow = floor(Rows/2);
pcol = floor(Cols/2);

    P = padarray(I,[prow,pcol]);

final = zeros(m,n);
for i = 1:m
```

```
cury = i+prow;
for j = 1:n
curx = j+pcol;
             Win = P((cury-prow):(cury+prow),(curx-pcol):(curx+pcol));
final(i,j) = length(find(Win<(P(cury,curx)-z)));
end
end

out = uint8(final);
norm = uint8(((final-min(min(final)))/(max(max(final)-
min(min(final))))).*255);

end
```

RGBtoSpect.m(convert RGB colour image to spectral representation)

```
function [Spect] = RGBtoSpect(R,G,B,All,mL,ML,nbins)

    %R = red value
    %G = Green value
    %B = Blue value
    %All = whiteSpectcyanSpectmagentaSpectyellow
    %SpectredSpectgreenSpectblueSpect

    %mL = minLambda (smallest wavelength to map)
    %ML = maxLambda (Largest wavelength to map)
    %nbins = number of bins

Spect = linspace(mL,ML,nbins);
%    ret = zeros(size(Spect));
%
%    if(R< = G && R< = B)
%        ret = ret + (R * All(:,1)');%All(:,1) = WhiteSpect
%        if(G< = B)
%            ret = ret + ((G-R)*All(:,2)');%All(:,2) = CyanSpect
%            ret = ret + ((B-G)*All(:,7)');%All(:,7) = BlueSpect
%      else
%            ret = ret + ((B-R)*All(:,2)');%All(:,2) = CyanSpect
%            ret = ret + ((G-B)*All(:,6)');%All(:,6) = GreenSpect
%      end
%    elseif(G< = R && G< = B)
%        ret = ret + (G * All(:,1)');%All(:,1) = WhiteSpect
%        if(R< = B)
%            ret = ret + ((R-G)*All(:,3)');%All(:,3) = MagentaSpect
%            ret = ret + ((B-R)*All(:,7)');%All(:,7) = BlueSpect
%      else
%            ret = ret + ((B-G)*All(:,3)');%All(:,3) = MagentaSpect
%            ret = ret + ((R-B)*All(:,5)');%All(:,5) = RedSpect
%      end
%    elseif(B< = R && B< = G)
%        ret = ret + (B * All(:,1)');%All(:,1) = WhiteSpect
%        if(R< = G)
%            ret = ret + ((R-B)*All(:,4)');%All(:,4) = YellowSpect
%            ret = ret + ((G-R)*All(:,6)');%All(:,6) = GreenSpect
%      else
%            ret = ret + ((G-B)*All(:,4)');%All(:,4) = YellowSpect
%            ret = ret + ((R-G)*All(:,5)');%All(:,5) = RedSpect
```

```
%         end
% end
%
% Spect = cat(1,Spect,ret);

%%%%% %%%%%%% %%%%%%% %%%%%Parallelization%%%%%% %%%%%%%%% %%%%%%% %%%%%%
ret = zeros(size(R,1),size(Spect,2));

    %%%
    IF1 = bsxfun(@and,bsxfun(@le,R,G),bsxfun(@le,R,B));
ind = find(IF1 = =1);
    Register = bsxfun(@times,R,All(:,1)');%All(:,1) = WhiteSpect;
    ret(ind,:) = ret(ind,:) + Register(ind,:);

    IF11 = bsxfun(@le,G,B);
    ind1 = find(IF11 = =1);
    Register = bsxfun(@times,(G-R),All(:,2)');%All(:,2) = CyanSpect
ret(ind1,:) = ret(ind1,:) + Register(ind1,:);
    Register = bsxfun(@times,(B-G),All(:,7)');%All(:,7) = BlueSpect
ret(ind1,:) = ret(ind1,:) + Register(ind1,:);

    ind2 = find(IF11 = =0);
    Register = bsxfun(@times,(B-R),All(:,2)');%All(:,2) = CyanSpect
ret(ind2,:) = ret(ind2,:) + Register(ind2,:);
    Register = bsxfun(@times,(G-B),All(:,6)');%All(:,6) = GreenSpect
ret(ind2,:) = ret(ind2,:) + Register(ind2,:);

    %%%
    IF1 = bsxfun(@and,bsxfun(@le,G,R),bsxfun(@le,G,B));
ind = find(IF1 = =1);
    Register = bsxfun(@times,G,All(:,1)');%All(:,1) = WhiteSpect
    ret(ind,:) = ret(ind,:) + Register(ind,:);

    IF11 = bsxfun(@le,R,B);
    ind1 = find(IF11 = =1);
    Register = bsxfun(@times,(R-G),All(:,3)');%All(:,3) = MagentaSpect
ret(ind1,:) = ret(ind1,:) + Register(ind1,:);
    Register = bsxfun(@times,(B-R),All(:,7)');%All(:,7) = BlueSpect
ret(ind1,:) = ret(ind1,:) + Register(ind1,:);

ind2 = find(IF11 = =0);
Register = bsxfun(@times,(B-G),All(:,3)');%All(:,3) = MagentaSpect
ret(ind2,:) = ret(ind2,:) + Register(ind2,:);
    Register = bsxfun(@times,(R-B),All(:,5)');%All(:,5) = RedSpect
ret(ind2,:) = ret(ind2,:) + Register(ind2,:);

    %%%
    IF1 = bsxfun(@and,bsxfun(@le,B,R),bsxfun(@le,B,G));
ind = find(IF1 = =1);
    Register = bsxfun(@times,B,All(:,1)');%All(:,1) = WhiteSpect
    ret(ind,:) = ret(ind,:) + Register(ind,:);

    IF11 = bsxfun(@le,R,G);
    ind1 = find(IF11 = =1);
    Register = bsxfun(@times,(R-B),All(:,4)');%All(:,4) = YellowSpect
ret(ind1,:) = ret(ind1,:) + Register(ind1,:);
    Register = bsxfun(@times,(G-R),All(:,6)');%All(:,6) = GreenSpect
ret(ind1,:) = ret(ind1,:) + Register(ind1,:);
```

```
    ind2 = find(IF11 = =0);
    Register = bsxfun(@times,(G-B),All(:,4)');%All(:,4) = YellowSpect
ret(ind2,:) = ret(ind2,:) + Register(ind2,:);
    Register = bsxfun(@times,(R-G),All(:,5)');%All(:,5) = RedSpect
ret(ind2,:) = ret(ind2,:) + Register(ind2,:);

Spect = cat(1,Spect,ret);
end
%%%%%%%%% %%%%%%%%% %%%%%%%%%CIF%%%%%%%% %%%%%%% %%%%%%% %%%%%%

function [cif] = CIF(I,CE,GT)

%I is original image
%CE is enhanced version of I
%GT is the ground truth

Les_Area_ind = find(GT(:,:,1) = =1);
Nles_Area_ind = find(GT(:,:,1) = =0);

if(size(CE,3)>1)
spe = rgb2gray(CE);
else
spe = CE;
end
ref = I(:,:,2);

Cspe = abs(mean(spe(Les_Area_ind))-mean(spe(Nles_Area_ind)));
Cref = abs(mean(ref(Les_Area_ind))-mean(ref(Nles_Area_ind)));

cif = Cspe/Cref;
end
%%%%%%%%%%%%%% %%%%%%%%%%%%% %%ICN%%%%%% %%%%%%%%%%%%% %%%%%%%%%
function [icn] = ICN(I,CE,GT)
%ICN Summary of this function goes here
% Detailed explanation goes here

%I is original image
%CE is enhanced version of I
%GT is the ground truth

Les_Area_ind = find(GT(:,:,1) = =1);
Nles_Area_ind = find(GT(:,:,1) = =0);

if(size(CE,3)>1)
spe = rgb2gray(CE);
else
spe = CE;
end

Cspe = abs(mean(spe(Les_Area_ind))-mean(spe(Nles_Area_ind)));

stdev = std(double(reshape(rgb2gray(I),1,size(I,1)*size(I,2))));

icn = stdev/Cspe;

end
```

References

1. Tan JKL. Current measures for the evaluation of acne severity. *Expert Review of Dermatology* 2008;3:595–603.
2. Fulton James J. Acne vulgaris. http://emedicine.medscape.com/article/ 1069804-overview. Last Accessed: November 16, 2013.
3. Tan JK, Tang J, Fung K, Gupta AK, Thomas DR, Sapra S, Lynde C, Poulin Y, Gulliver W, Sebaldt RJ. Prevalence and severity of facial and truncal acne in a referral cohort. *Journal of Drugs in Dermatology* 2008;7:551–556.
4. Cordin L, Linderberg S, Hurtado M, Hill K, Eaton S. Acne vulgaris: a disease of Western civilization. *Archives of Dermatology* 2002;138:1584–1590.
5. Savage LJ, Layton AM. Treating acne vulgaris: systemic, local and combination therapy. *Expert Review of Clinical Pharmacology* 2010;3:563–580.
6. Simpson NB, Cunliffe WJ. Disorders of the sebaceous glands. In Burns T, Breathnach S, Cox N, Griffiths C, eds., *Rook's textbook of dermatology*, 7th ed. Malden, MA: Blackwell, 2004:43.1–43.75.
7. Lehmann HP, Robinson KA, Andrews JS, Holloway V, Goodman SN. Acne therapy: a methodologic review. *Journal of the American Academy of Dermatology* 2002;47:231–240.
8. Adityan B, Kumari R, Thappa DM. Scoring systems in acne vulgaris. *Indian Journal of Dermatology, Venereology, and Leprology* 2009;75:323–326.
9. Pillsbury DM, Shelley WB, Kligman AM. *Dermatology*. Philadelphia: WB Saunders, 1956.
10. Witkowski JA, Parish LC. The assessment of acne: an evaluation of grading and lesion counting in the measurement of acne. *Journal of the American Academy of Dermatology* 2004;22:394–397.
11. Cook CH, Centner RL, Michaels SE. An acne grading method using photographic standards. *Archives of Dermatology* 1979;115:571-575.
12. Lucchina L, Kollias N, Gillies R, Phillips SB, Muccini JA, Stiller MJ, Trancik RJ, Drake LA. Fluorescence photography in the evaluation of acne. *Journal of the American Academy of Dermatology* 1996;35:58–63.
13. Lucky A, Barber B, Girman C, William J, Ratterman J, Waldstreicher J. A multi-rater validation study to assess the reliability of acne lesion counting. *Journal of the American Academy of Dermatology* 1996;35:559–565.
14. Doshi A, Zaheer A, Stiller MJ. A comparison of current acne grading systems and proposal of a novel system. *International Journal of Dermatology* 1997;36:416–418.
15. O'Brien SC, Lewis JB, Cunliffe WJ. The Leeds revised acne grading system. *Journal of Dermatological Treatment* 1998;44:215–220.
16. Hayashi N, Akamatsu H, Kawashima M. Establishment of grading criteria for acne severity. *Journal of Dermatology* 2008;35:255–260.
17. Eichenfield LF, Nighland M, Rossi AB, Cook-Bolden F, Grimes P, Fried R, Levy S. Phase 4 study to assess tretinoin pump for the treatment of facial acne. *Journal of Drugs in Dermatology* 2008;7:1129–1136.
18. Law MP, Chuh AA, Lee A, Molinari N. Acne prevalence and beyond: acne disability and its predictive factors among Chinese late adolescents in Hong Kong. *Clinical and Experimental Dermatology* 2010;35:16–21.

19. Naieni FF, Akrami H. Comparison of three different regimens of oral azithromycin in the treatment of acne vulgaris. *Indian Journal of Dermatology* 2006;51:255–257.

20. El-Akawi Z, Abdel-Latif Nemr N, Abdul-Razzak K, Al-Aboosi M. Factors believed by Jordanian acne patients to affect their acne condition. *Eastern Mediterranean Health Journal* 2006;12:840–846.

21. Do JE, Cho S-M, In SI, Lim KY, Lee S, Lee E-S. Psychosocial aspects of acne vulgaris: a community-based study with Korean adolescents. *Annals of Dermatology* 2009;21:125–129.

22. Hanisah A, Khairani O, Shah SA. Prevalence of acne and its impact on the quality of life in school-aged adolescents in Malaysia. *Journal of Primary Healthcare* 2009;1:20–25.

23. Kokandi A. Evaluation of acne quality of life does not correlate with severity of disease. *Dermatology Research and Practice* 2010;2010:410809.

24. Demircay Z, Kus S, Sur HD. Predictive factors for acne flare during isotretinoin treatment. *European Journal of Dermatology* 2008;18:452–456.

25. Charakida A, Charakida M, Chu AC. Double-blind, randomized, placebo-controlled study of a lotion containing triethyl citrate and ethyl linoleate in the treatment of acne vulgaris. *British Journal of Dermatology* 2007;157:569–574.

26. U.S. Food and Drug Administration. Guidance for industry. Acne vulgaris: developing drugs for treatment. Rockville, MD: U.S. Food and Drug Administration, 2005. http://www.fda.gov/downloads/Drugs/.../Guidances/UCM071292.pdf

27. Phillips SB, Kollias N, Gillies R, Muccini JA, Drake LA. Polarized light photography enhances visualization of inflammatory lesions of acne vulgaris. *Journal of the American Academy of Dermatology* 1997;37:948–952.

28. Rizova E, Kligman A. New photographic techniques for clinical evaluation of acne. *Journal of the European Academy of Dermatology and Venereology* 2001;15(Suppl 3):13–18.

29. Do TT, Zarkhin S, Orringer JS, Nemeth S, Hamilton T, Sachs D, Voorhees JJ, Kang S. Computer-assisted alignment and tracking of acne lesions indicate that most inflammatory lesions arise from comedones and de novo. *Journal of the American Academy of Dermatology* 2008;58:603–608.

30. Fujii H, Yanagisawa T, Mitsui M, Murakami Y, Yamaguchi M, Ohyama N, Abe T, Yokoi I, Matsuoka Y, Kubota Y. Extraction of acne lesion in acne patients from Multispectral Images. *30th Annual International Conference of the IEEE Engineering in Medicine and Biology Society*. Washington, DC: IEEE, 2008:4078-4081.

31. Bae Y, Nelson JS, Jung B. Multimodal facial colour imaging modality for objective analysis of skin lesions. *Journal of Biomedical Optics* 2008;13:1–19.

32. Freid R, and Wechsler A. Psychological problems in the acne patient. *Dermatologic Therapy* 2006;19:237–240.

33. Cotterill JA, Cunliffe WJ. Suicide in dermatological patients. *British Journal of Dermatology* 1997;137:246–250.

34. *Clinical practice guidelines: management of acne.* MOH/PAK/234.12(GU). Putrajaya, Malaysia: Ministry of Health Malaysia, January 2012.

35. Mukherjee J. Local rank transform: properties and applications. *Pattern Recognition Letters* 2011;32:1001–1008.

36. Gillespie AR. Enhancement of multispectral thermal infrared images: decorrelation contrast stretching. *Remote Sensing of Environment* 1992;42:147–155.
37. Wikipedia. Decorrelation. http://en.wikipedia.org/w/index.php?title=Decorrelation&oldid=512403761
38. Smits B. An RGB to spectrum conversion for reflectances. *Journal of Graphics Tools* 1999;4:11–22.

Index